Exploring Expressions and Equations

with THE GEOMETER'S SKETCHPAD VERSION 5

Addressing the Common Core State Standards for Mathematics

Key Curriculum Press
INNOVATORS IN MATHEMATICS EDUCATION

Writers:	Dan Bennett, Christopher Casey, Greg Clarke, Larry Copes, Deidre Grevious, Lynn Hughes, Rhea Irvine, Ross Isenegger, Nick Jackiw, Tobias Jaw, Amy Lamb, Paul Kunkel, Ann Lawrence, Andres Marti, Daniel Scher, Nathalie Sinclair, Scott Steketee, Kelly Stewart, Kevin Thompson
Reviewers:	Karen Anders, Susan Beal, Janet Beissinger, Dudley Brooks, Greg Clarke, Gord Cooke, Larry Copes, Judy Dussiaume, Paul Gautreau, Shawn Godin, Paul Goldenberg, Lynn Hughes, Scott Immel, Ross Isenegger, Sarah Kasten, Cathy Kelso, Amy Lamb, Dan Lufkin, Aaron Madrigal, Linda Modica, Margo Nanny, Henry Picciotto, Nicolle Rosenblatt, Joan Scher, Dick Stanley, Tom Steinke, Glenda Stewart, John Threlkeld, Philip Wagreich, Ken Waller, Bill Zahner, Danny Zhu
Field Testers:	Laura Adler, Joëlle Auberson, Vera Balarin, Kim Beames, Judy Bieze, Caron Cesa, Heather Darby, Robin Glass, Cherish Hansen, Layne Hudes, Lynn Hughes, Scott A. Immel, Susan Friedman, Ann Lawrence, Vanessa Mamikunian, Michelle Mancini, Michelle Moore, Margo Nanny, Kelly O'Keefe, Tina Pierorazio, Leslie Profeta, Dechelle Rasheed, Cheryl Schafer, Joan Scher, Kimberly Scheier, JoAnne Searle, Char Soucy, Jessie Starr, Ruth Steinberg, Nancy Stevenson, Terry Suetterlein, Mona Sussman, William Vaughn, Ethan Weker, Angie Whaley
Editors:	Andres Marti, Elizabeth DeCarli
Contributing Editors:	Rhea Irvine, Daniel Scher, Josephine Noah, Scott Steketee, Cindy Clements, Joan Lewis, Silvia Llamas-Flores, Kendra Lockman, Lenore Parens, Glenda Stewart, Kelly Stewart
Editorial Assistant:	Tamar Chestnut
Production Director:	Christine Osborne
Production Editors:	Angela Chen, Andrew Jones, Christine Osborne
Other Contributers:	Judy Anderson, Elizabeth Ball, Tamar Chestnut, Brady Golden, Ashley Kuhre, Nina Mamikunian, Marilyn Perry, Emily Reed, Ann Rothenbuhler, Juliana Tringali, Jeff Williams
Copyeditor:	Jill Pellarin
Cover Designer:	Anderson-Carey Design
Cover Photo Credit:	© Alistair Berg
Printer:	Lightning Source, Inc.
Executive Editor:	Josephine Noah
Publisher:	Steven Rasmussen

Key Curriculum Press
1150 65th Street
Emeryville, CA 94608
510-595-7000
editorial@keypress.com
www.keypress.com

ISBN: 978-1-60440-226-1
10 9 8 7 6 5 4 3 2 1   15 14 13 12 11

Contents

Chapter 6: Equations and Inequalities

Chapter 7: Polynomials

Exploring Expressions and Equations in Grades 6–8 with The Geometer's Sketchpad
© 2012 Key Curriculum Press

Downloading Sketchpad Documents

Getting Started

All Sketchpad documents (sketches) for *Exploring Expressions and Equations in Grades 6–8 with The Geometer's Sketchpad* are available online for download.

- Go to www.keypress.com/gsp5modules.
- Log in using your Key Online account, or create a new account and log in.
- Enter this access code: EE68w361
- A Download Files button will appear. Click to download a compressed (.zip) folder of all sketches for this book.

The downloadable folder contains all of the sketches you need for this book, organized by chapter and activity. The sketches require The Geometer's Sketchpad Version 5 software to open. Go to www.keypress.com/gsp/order to purchase Sketchpad, or download a trial version from www.keypress.com/gsp/download.

Types of Sketches

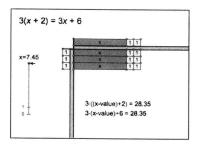

Student Sketches: In most activities, students use a prepared sketch that provides a model, a simulation, or a complicated construction to investigate relationships. You will often use the student sketch to introduce an activity, guided by the Activity Notes. The name of a Student Sketch usually matches the activity title, and is referenced on the Student Worksheet for the activity, such as **Broccoli & Brussels Sprouts.gsp.**

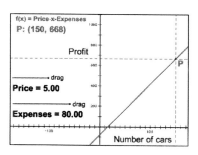

Presentation Sketches: Some activities include sketches designed for use with a projector or interactive whiteboard, either for a teacher presentation or a whole-class activity. Presentation Sketches often have action buttons to enhance your presentation. For activities in which students create their own constructions, the Presentation Sketch can be used to speed up, summarize, or review the mathematical ideas from the activity. The name of a Presentation Sketch always ends with the word "Present," such as **Car Wash Present.gsp.**

Sketchpad Resources

Sketchpad Learning Center

The Learning Center provides a variety of resources to help you learn how to use Sketchpad, including overview and classroom videos, tutorials, Sketchpad Tips, sample activities, and links to online resources. You can access the Learning Center through Sketchpad's start-up screen or through the Help menu.

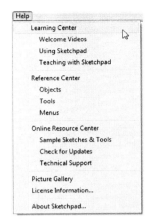

The Learning Center has three main sections:

Welcome Videos

These videos introduce Sketchpad from the point of view of students and teachers, and give an overview of the big ideas and new features of Sketchpad 5.

Using Sketchpad

This section includes 12 self-guided tutorials with embedded videos, 70 Sketchpad Tips, and links to local and online resources.

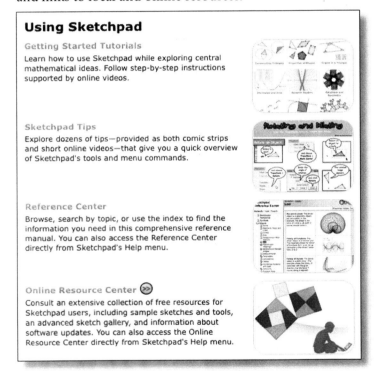

Teaching with Sketchpad

This section includes videos and articles describing how teachers make effective use of Sketchpad and how it affects their students' attitudes and mathematical understanding. There are over 40 sample activities, each with an overview, teaching notes, student worksheet, and sketches, that you can use with students to support your grade level and curriculum.

Other Sketchpad Resources

Exploring Expressions and Equations in Grades 6–8 with The Geometer's Sketchpad activities are an excellent introduction to Sketchpad for both students and teachers. If you want to learn more, Sketchpad contains resources for beginning and advanced users.

- **Reference Center:** This digital resource, which is accessed through the Help menu, is the complete reference manual for Sketchpad, with detailed information on every object, tool, and menu command. The Reference Center includes a number of How-To sections, an index, and full-text search capability.

- **Online Resource Center:** The Geometer's Sketchpad Resource Center (www.dynamicgeometry.com) contains many sample sketches and advanced toolkits, links to other Sketchpad sites, technical information (including updates and frequently asked questions), and detailed documentation for JavaSketchpad, which allows you to embed dynamic constructions in a web page.

- **Sketch Exchange:** The Sketchpad Sketch Exchange™ (sketchexchange.keypress.com) is a community site where teachers share sketches and other resources with Sketchpad users. Browse by keyword or topic for sketches that interest you, or ask questions and share ideas in the forum.

- **Sample Sketches & Tools:** You can access many sketches, including some with custom tools, through Sketchpad's Help menu. You can use some sample sketches as demonstrations, others to get tips and information about particular constructions, and others to access custom tools that you can use to perform special constructions. These sketches are also available under General Resources at the Sketchpad Resource Center (www.dynamicgeometry.com).

- **Sketchpad LessonLink™:** This online subscription service includes a library of more than 500 prepared activities (including those used in this book) aligned to leading math textbooks and state standards for grades 3–12. For more information, additional sample activities, or a trial subscription, go to www.keypress.com/sll.

- **Online Courses:** Key Curriculum Press offers moderated online courses that last six weeks, allowing you to immerse yourself in learning how to use Sketchpad in your teaching. For more information, see Sketchpad's Learning Center, or go to www.keypress.com/onlinecourses.

- **Other Professional Development:** Key Curriculum Press offers free webinars on a regular basis. You can also arrange for one-day or three-day face-to-face workshops for your district or school. For more information, go to www.keypress.com/pd.

Addressing the Common Core State Standards

The Common Core State Standards emphasize the importance of students gaining expertise with a variety of mathematical tools, including dynamic geometry® software such as The Geometer's Sketchpad. The Standards for Mathematical Practice specify that students should use technological tools to explore and better understand mathematical concepts.

5. Use appropriate tools strategically.

Mathematically proficient students consider the available tools when solving a mathematical problem. These tools might include pencil and paper, concrete models, a ruler, a protractor, a calculator, a spreadsheet, a computer algebra system, a statistical package, or dynamic geometry software. Proficient students are sufficiently familiar with tools appropriate for their grade or course to make sound decisions about when each of these tools might be helpful, recognizing both the insight to be gained and their limitations. They are able to use technological tools to explore and deepen their understanding of concepts. (Common Core State Standards for Mathematics, 2010, www.corestandards.org)

This collection of Sketchpad activities supports you in addressing the Common Core State Standards for Mathematics for Grades 6 to 8. In addition to supporting the Standards for Mathematical Practice, these activities give students the opportunity to explore content from the domain of Expressions and Equations, as well as Functions in Grade 8. As a set, they address the standard cluster statements listed below.

Grade 6: Expressions and Equations

- Apply and extend previous understandings of arithmetic to algebraic expressions.
- Reason about and solve one-variable equations and inequalities.
- Represent and analyze quantitative relationships between dependent and independent variables.

Grade 7: Expressions and Equations

- Use properties of operations to generate equivalent expressions.
- Solve real-life and mathematical problems using numerical and algebraic expressions and equations.

Grade 8: Expressions and Equations

- Work with radicals and integer exponents.
- Understand the connections between proportional relationships, lines, and linear equations.
- Analyze and solve linear equations and pairs of simultaneous linear equations.

Grade 8: Functions

- Define, evaluate, and compare functions.
- Use functions to model relationships between quantities

Overview of Content

The content of each chapter is described below, as well as how the activities in that chapter correlate to the Common Core State Standards. All of these activities strongly correlate to the Standards for Mathematical Practice. The Geometer's Sketchpad is a mathematical tool that students can use strategically, allows them to model with mathematics, and motivates them to make sense of problems and persevere in solving them. In addition, the worksheets and activity notes are designed to help students reason abstractly and quantitatively, look for and make use of structure, and look for and express regularity in repeated reasoning.

Chapter 1: Expressions

Students use models that build the concept of a variable. Starting with objects that represent numerical values, students think algebraically as they use problem-solving skills and begin to internalize the idea of equivalence. Additional models allow them to relate equivalent expressions to different ways of representing the same geometric situation.

Chapter 2: Operations

Students build equivalent expressions and explore properties of operations with models such as dynamic algebra tiles and algebars. The dynamic nature of Sketchpad allows students to vary the value of variables to check for equivalence both visually and by calculation. Students also explore properties of exponents, working with both positive and negative exponents, and developing and expressing the multiplication, division, and power laws of exponents.

Chapter 3: Functions

Function machines provide an engaging introduction to the concepts of inputs, outputs, and functions, and give students practice identifying and expressing number patterns. Students informally explore the inverse relationship between addition and subtraction and between multiplication and division, and they write equations for linear functions.

Chapter 4: Graphs of Linear Functions

Students move among multiple representations, including graphs, tables, and verbal descriptions, to explore linear graphs and systems. Because graphs, tables, and equations are linked, students see how a change in one representation relates to a change in another. Students explore time-distance graphs and time-speed graphs and use linear equations to model relationships between quantities and solve real-world problems.

Chapter 5: Nonlinear Functions

Students explore both linear and nonlinear functions, developing fluency moving among symbolic, graphical, and tabular representations. Students compare the growth of linear, quadratic, and exponential functions.

Chapter 6: Equations and Inequalities

Students work in a balancing environment to solve linear equations, and also use the undoing approach, which emphasizes the order of operations and inverse operations. Students reason about inequalities and solve and graph the solutions to one-variable inequalities and linear inequalities.

Chapter 7: Polynomials

Using dynamic algebra tiles, students evaluate polynomial expressions, multiply polynomials, and factor polynomials. In the final activity, students explore the relationship among the symbolic, graphical, and rectangular area models for a quadratic function.

1

Expressions

Circles and Squares: Representing an Unknown

Students manipulate a dynamic model that uses symbols to represent numerical values. Students develop early algebraic reasoning skills by writing statements, exploring the meaning of the equal sign, writing equivalent addition statements, and solving for an unknown. This activity is led by the teacher; there is no Student Worksheet.

Circles and Squares: Two Unknowns

Students use algebraic thinking as they work with addition statements in which the addends are represented by two symbols. By analyzing the information from two or more statements, students deduce the numerical values of the symbols. This activity is lead by the teacher.

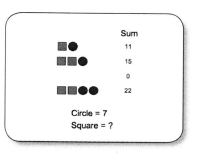

Circles, Squares, and Triangles: Solving for Three Unknowns

Students use algebraic thinking as they work with addition statements in which the addends are represented by three symbols. By analyzing the information from multiple addition statements, students deduce the numerical values of the symbols. Students' intuitive solution methods lay a foundation for their study of algebra.

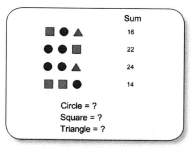

Odd Squares: Adding Consecutive Odd Numbers

Students see an animated representation and discover a geometric relationship that allows them to find the sum of consecutive odd numbers without adding the individual terms. They generalize their findings by writing an algebraic expression for finding the sum of the first n odd numbers for any value of n.

Signed Tiles: Evaluating Expressions

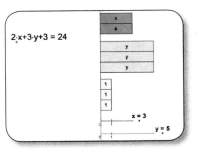

Students use dynamic algebra tiles to represent expressions with one or two variables. They evaluate the expressions using paper and pencil, and then check their answers by manipulating the tiles and performing Sketchpad calculations.

Pool Border: Equivalent Expressions

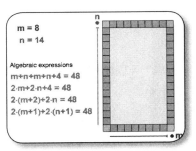

Students use geometric models to develop algebraic expressions for determining the number of unit tiles needed to surround a pool. As they change the size of the pool, they see that the algebraic expressions are equivalent. Students also learn how to convert between equivalent expressions by distributing and combining like terms.

Circles and Squares: Representing an Unknown

ACTIVITY NOTES

INTRODUCE

Project the sketch for viewing by the class. Expect to spend about 15 minutes.

1. Open **Circles Squares An Unknown.gsp** and go to page "Introduction." Check that the values of the square and circle are 4 and 1, respectively. To change a value, double-click it, enter a new value in the dialog box that appears, and click OK. Ask students to describe what they see. The left side of the screen contains eight rows of squares and circles. A vertical line separates the objects from a column of zeroes. The bottom of the screen shows the numerical values of the square and circle.

   Explain, *We can drag any row of squares and circles that we want across the line. When I drag the first row, the computer will add the values of the square and circle.*

   Drag the square and circle across the divider line. The 0 will change to a 5, indicating the sum of 4 and 1.

2. Distribute paper. *I'd like each of you to record the information about the sum of the square and circle on your paper. There is more than one correct answer. You decide how to represent the information.* Ask volunteers to share what they wrote. Here are possible responses.

   5

   4 + 1 = 5

   ▢ ● 5

   ▢ ● = 5

   ▢ + ● = 5

3. If students do not offer the last statement above, ask them to write a statement that includes the square, the circle, an addition sign, and an equal sign. Read the statement together as a class: "Square plus circle equals 5." Drag the row back to the left of the line.

DEVELOP

Continue to project the sketch. Expect to spend about 30 minutes.

4. Focus on writing addition statements with symbols. Pick another row of circles and squares. Ask students to compute the sum of the symbols (before they are dragged across the vertical line) and write a corresponding statement. Ask volunteers to share their statements. Drag the row across the vertical line to check.

5. Write this statement on the board.

$$\square + \square + \bullet = 9$$

Below it, write, "9 = ." Ask students to complete the sentence so that it gives the same information as the original statement. They might write,

$$9 = \square + \square + \bullet$$

6. The order of the symbols can be changed without changing the result. Show the statement above and ask students to write another true statement using the same symbols in a different order. Here are two.

$$9 = \square + \bullet + \square$$
$$9 = \bullet + \square + \square$$

7. Return to the sketch. Pick other rows of symbols and ask students to write equivalent number statements. To add variety to the problems, press *New Values* to change the numerical values of the circle and the square.

Finding the Value of an Unknown Symbol

8. Go to page "Find the Unknown." Explain that the model is like the previous one, but now the value of the square is hidden. The goal is to figure out its value.

9. Drag the first row of symbols across the vertical line. For the sake of example, let's say that the value of the circle is shown as 7 and the reported sum is 10.

Ask students to write the addition statement shown on their papers. Have a volunteer share the statement on the board.

Is it possible for us to figure out the value of the square if we know the sum of the square and the circle? Students may suggest that one way to solve the problem is to replace the circle with its numerical value.

$$\square + 7 = 10$$

Students might solve this problem by phrasing it, for example, *What number plus 7 equals 10?* When the class has solved the problem, press *Show Answer* to reveal the value of the square.

10. Create new problems by pressing *New Challenge*.

11. Increase the challenge by dragging either of the following rows. In both cases, the circles must be added in order to determine the value of the square.

12. For an added level of difficulty, drag a row containing two or more squares. If, for example, 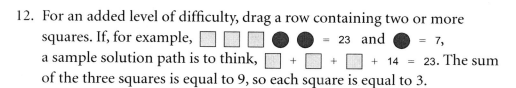 a sample solution path is to think, □ + □ + □ + 14 = 23. The sum of the three squares is equal to 9, so each square is equal to 3.

SUMMARIZE

Working away from computers, expect to spend about 15 minutes.

13. Have students write an explanation for someone who has never used the "circles and squares" model, explaining how to find the hidden value of the square. Students can include pictures and examples.

EXTEND

1. For students who would benefit from more individualized work, provide an opportunity to use the student sketch in pairs at a later time.

2. Pairs of students can create challenges for each other by changing the values of the symbols on page "Make Your Own." Now students can enter values for the square and circle that either exceed 8 or are negative; decimal values are rounded to the nearest whole number.

Circles and Squares: Two Unknowns

INTRODUCE

Project the sketch for viewing by the class. Expect to spend about 30 minutes.

1. Introduce the Sketchpad model. Open **Circles Squares Two Unknowns.gsp** and go to page "Game 1." *We've got a problem to solve. There are rows of squares and circles. The squares all have a hidden value, a whole number 0 through 8. Every square has the same value. The circles also have a whole number value 0 through 8. We don't know what the value of the square is or what the value of the circle is—that's what we need to figure out!*

   We can drag any row of squares and circles across the screen. When I drag a row across the line, the computer will add together the value of the squares and circles. Drag the first row across the divider line. The numeral 0 will change to 9, indicating the sum of the two numbers. With the aid of the class, write this sum as

2. ***Do we now know the values of the circle and square?*** Take responses. Students will propose that there are many combinations of two numbers that add up to 9. Ask students to make a list of these values. Remind students that neither symbol is greater than 8.

   Ask volunteers to share square/circle values from their lists. Write these on the board in an organized list.

□	●
1	8
2	7
3	6
4	5
5	4
6	3
7	2
8	1

3. ***How do you think we can figure out the actual values of the square and circle?*** Let students suggest the idea of dragging another row across the divider line. Allow them to choose which row to drag. In the example that follows, we assume students pick the second row.

   • Drag the second row □ □ ● to the right of the divider. A sum of 11 appears.

Exploring Expressions and Equations in Grades 6–8 with The Geometer's Sketchpad
© 2012 Key Curriculum Press

- *How can this new information help us?* Give students time to check their list of square/circle values to determine which pair of numbers satisfies the new statement.

- Make a new column on the board and label it ⬜⬜⚫. In this column, ask the class to write the sum of each pair of numbers. Only one pair gives the desired sum: square = 2 and circle = 7.

⬜ ⚫		⬜⬜⚫
1	8	10
2	7	11
3	6	12
4	5	13
5	4	14
6	3	15
7	2	16
8	1	17

- Press *Show Answers* to confirm the values of the two symbols. Drag the first two rows of symbols back to the left of the divider.

- *We solved this problem by dragging over the first and second rows of symbols. Suppose instead we drag over the third and fourth rows of symbols. If we solve the problem with that information, will we get the same answer?* Some students may realize that the answer will be the same regardless of which rows are picked. Other students may not be sure. Drag over the third and fourth rows, and give students time to work with the results before discussing as a class.

DEVELOP

Continue to project the sketch. Expect to spend about 30 minutes.

4. Go to page "Game 2." Drag over the first row of symbols. The sum is 26. *In this game, the range of possible values for the square and circle has changed. Now, each symbol represents a whole number from 0 through 20. What should we do first to solve this problem?*

Some students are likely to suggest making a list again. With the class, list all the square and circle combinations. There are 15 possible pairs, starting with square = 6, circle = 20 and ending with square = 20, circle = 6. (Note that it is possible that the values of square and circle are the same.) As before, this approach will allow students to solve the problem. The list, however, is quite long, prompting the question of whether there's a more efficient way to solve the problem.

5. ***Our list is really long. Can you think of another way to solve the problem that might be faster?*** Ask for permission to erase the list, and then allow plenty of time for students to consider this question. Some may see a strategy more quickly than others.

6. The steps that follow describe a way to guide discovery of this strategy in a way that includes writing addition statements. Some students will find this way of working helpful in continuing to solve problems. Judge for yourself how much you want to guide students. Alternatively, invite a volunteer to the computer to demonstrate the ideas expressed previously and in the next steps.

   • Drag the second row of symbols (the two squares and circle) across the divider line. Their sum is 41. ***Now that we have two rows of information, I'd like you to write each row as a number sentence using circles, squares, and plus and equal signs.***

   • Ask a volunteer to write the statements on the board.

$$\square + \bullet = 26$$
$$\square + \square + \bullet = 41$$

   • ***What do these statements have in common and what's different about them?*** Elicit from students that each statement contains a square and a circle, and that the second statement contains an extra square.

   • Draw a box around the square and circle in the second statement and ask, ***What do you know about this sum?*** From the first statement, students know that the sum is 26. Let students suggest replacing the two symbols with 26.

$$\square + \boxed{\square + \bullet} = 41$$
$$\square + \quad 26 \quad = 41$$

   • ***Now can you figure out the value of the square?*** One way to think about it is to ask what number plus 26 equals 41. [The square is equal to 15.]

   • ***Now that you know the value of the square, can you figure out the value of the circle?*** [Because square + circle = 26, and the square is equal to 15, the circle is equal to 11.]

   • Check the answer to the problem by pressing *Show Answers*.

Solving for two unknown values in two statements is an important concept that students will encounter again in algebra. These statements are known as *simultaneous equations* in algebra.

Solving More Problems

7. Create a new problem by pressing *New Challenge*. Drag the following two rows of symbols over the line.

In this case, students can replace the square and circle in the second statement with a number. They'll be left with a statement that gives the sum of two circles. If students determine, for instance, that circle + circle = 10, then circle = 5.

> Some students will quickly realize that the values of the circles and squares can always be found by dragging the first two rows—and that those rows are the easiest to use. Make sure they explore other pairs of rows.

8. Continue creating new problems. For each problem, press *New Challenge* and solve by selecting different rows of symbols. Explain that by choosing different pairs of rows each time, students can not only take on new challenges, but also look for a good general strategy for finding the unknowns.

Some good pairs of rows to choose follow. Present several of these pairs so that students have the opportunity to make a conjecture about why these pairs are useful. Alternatively, you may let the class decide on the rows to drag and observe whether some students suggest pairs like these.

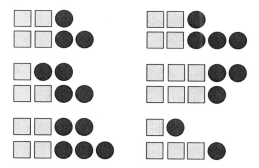

SUMMARIZE

> Continue to project the sketch. Expect to spend about 30 minutes.

9. Lead a discussion in which students develop and describe the general strategy for solving for the unknowns. Remind the class of the question that was posed earlier: ***Is there a way to figure out the values of the square and circle that is faster than making a list?***

If you chose the pairs of rows as the class solved more problems, ask, ***For each problem you've solved, I picked two rows of symbols. How do you think I chose those rows? Was there something nice about the rows we used? Was there something the same about all the pairs of rows?***

If students chose the pairs of rows, ask, *I noticed that some of you figured out pairs of rows that were easy to work with. How did you choose those rows? Was there something nice about those rows? Was there something the same about all the pairs you chose?*

The discussion should elicit these ideas.

- Good pairs of rows differed by one or more squares, or by one or more circles (and not by both squares and circles).

- Good pairs of rows had some symbols in common.

- By replacing the shared symbols with their numerical value, it was easy to find the value of the "extra" symbol or symbols.

10. *Can you think of a pair of rows that would make it really hard for us to solve the problem?* Take responses and try one pair with the sketch. A harder-to-solve pair of rows might be the seventh and eighth rows. It's still possible to figure out the values of the two symbols, but not as easy.

Creating the Problems

11. Explain that students will each create a problem, on paper, for a classmate to solve. Have students brainstorm ideas for presenting their problems.

- Will they draw many rows of symbols or just two rows? For simplicity, it is probably easier if students show just two rows. (Students may use rows from the sketch or create their own combinations of symbols.)

- How will the answer to the problem be shown? Will it be written on a separate sheet of paper or on the same sheet, perhaps hidden behind a folded edge?

Students should create their problems on their own, creating the rows of symbols and the values of the symbols. For variety's sake, students can draw symbols other than the square and circle.

Students should exchange problems with a classmate and observe as their partners solve the problems.

 ACTIVITY NOTES

EXTEND

1. For students who would benefit from more individualized work with solving for unknowns, provide opportunities to use pages "Game 1" or "Game 2" at a later time.

2. Pairs of students can create challenges for each other by changing the values of the symbols on page "Make Your Own" of **Circles Squares Two Unknowns.gsp.** As one student looks away, the other student double-clicks the value of the circle and then the square, changing each number to a new value. Pressing *Hide Answers* conceals the values of the two symbols. The sketch is now ready for use by the other student. When she is convinced that she knows the value of the two symbols, she should explain her reasoning to her partner and then press *Show Answers* to check her work.

   Let students know that to change the value of the square or circle on page "Make Your Own," they will double-click the number with the **Arrow** tool. In the dialog box that appears, they will enter a new number in the value field and click **OK.**

Circles, Squares, and Triangles: Solving for Three Unknowns

INTRODUCE

Project the sketch on a large-screen display for viewing by the class. Expect to spend about 5 minutes.

1. Open **Circles Squares Three Unknowns.gsp.** Go to page "Three Unknowns" and introduce the model. Explain, *We see rows of circles, squares, and triangles here. The squares all have a secret value, a number between 0 and 10. Every square has the same value, but we don't know what it is. That's what we need to figure out! The circles and triangles also have values that are between 0 and 10. We need to find their values too.*

   We can drag any row of symbols across the vertical line. The computer will add the value of the symbols.

2. Drag the first row across the vertical line. Record the information as

   Do we have enough information to know the value of each symbol? Let students come to an agreement that there is not enough information. *It will be up to you to find the value of each symbol when you work at the computers. You can drag any row you want across the screen to find the sum of its symbols. Drag as many rows as you need. There are many ways to solve this problem, so try different ways. If not every method works, that's fine.*

DEVELOP

Expect students at computers to spend about 25 minutes.

3. Assign students to work at computers and tell them where to locate **Circles Squares Three Unknowns.gsp.**

4. Distribute the worksheet. Tell students to work through worksheet steps 1–5 and do the Explore More if they have time. Let students know that sheets of blank paper are available if they need more room to record their work. Encourage students to ask a neighbor for help if they have a question about using Sketchpad.

To help students who make some progress but become stuck, use the notes at right as a guide to helping them persevere in doing their own thinking. You may also bring the class together for a check-in, asking students to share their strategies.

5. Let pairs work at their own pace. As you circulate, note the strategies that students use. Steps 6–9 describe four possible approaches.

6. Using guess and check, students begin by concentrating on a single row of symbols and its sum. Through either a systematic or random approach, they find sets of numerical values that satisfy the equation. For each set, students drag other rows to see whether the values work for them as well.

EXAMPLE

Row 2 tells us that ● + ● + ▢ = 22. By being systematic, we can list all the values of the circle and square that satisfy the equation.

circle = 6 and square = 10

circle = 7 and square = 8

circle = 8 and square = 6

circle = 9 and square = 4

circle = 10 and square = 2

Only one pair of values (circle = 10 and square = 2) also satisfies the row 4 equation ▢ + ▢ + ● = 14. Substituting these numerical values into the row 1 equation ▢ + ● + ▲ = 16 and solving for the value of the triangle reveals that triangle = 4.

> Remind students that the values of the circle and square are each less than or equal to 10. Thus, circle = 1 and square = 20 is not an option.

NOTES

Most students will probably try this approach first. Help students to be systematic in listing the values that satisfy their equations. Those students who pick the first row of symbols will find it more challenging to make their lists than those students who pick one of the other rows: These rows contain only two of the symbols.

7. Looking for what's the same and subtracting rows, students compare rows to see which symbols they have in common. If two rows share some of the same symbols, students "cancel out" these symbols by subtracting one row from another.

EXAMPLE

Looking at rows 2 and 4 we see

● + ● + ▢ = 22

▢ + ▢ + ● = 14

Both rows contain a circle and a square. When we subtract one equation from the other, we cancel out one circle and one square to get

● − ▢ = 8

> Notice that we're using a guess-and-check approach here.

Because the values of all symbols are between 0 and 10, there are only three possible pairs of numbers that give a difference of 8 (circle = 10, square = 2; circle = 9, square = 1; circle = 8, square = 0). Testing these pairs in either equation reveals that circle = 10 and square = 2.

Substituting these values into the row 1 equation allows us to solve for the unknown value of the triangle!

NOTES

There are other rows students may choose that share common symbols. Help students realize that regardless of the rows they pick, the strategy of subtracting one row from another is the same.

8. Looking for what's the same and adding rows, students find two rows with the same symbols, in different numbers. They combine the two rows by adding one to the other. Using this new equation, students gain useful information about the two symbols.

EXAMPLE

Rows 3 and 7 tell us that

● + ● + ▲ = 24   and   ▲ + ▲ + ● = 18

These two equations can be added together to get

● + ● + ● + ▲ + ▲ + ▲ = 42

Because three circles plus three triangles equals 42, that means

● + ▲ = 14

This information can be substituted into row 1, ▢ + ● + ▲ = 16, to obtain square = 2.

NOTES

The most challenging aspect of this method is simplifying the equation that is the result of adding two rows. Help some students by rewriting it.

● + ● + ● + ▲ + ▲ + ▲ = 42

(● + ▲) + (● + ▲) + (● + ▲) = 14 + 14 + 14

Check that students can use this new equation to explain why

● + ▲ = 14

Note too that there are other rows students may pick that share the same two symbols. Help students to see that regardless of the rows they pick, the strategy of adding one row to another is the same.

Exploring Expressions and Equations in Grades 6–8 with The Geometer's Sketchpad
© 2012 Key Curriculum Press

9. Substituting one symbol for another, students find a way to rewrite one symbol in terms of another symbol. They then substitute the new expression they've obtained into another row.

    EXAMPLE

    By looking for what's the same and subtracting rows, students determine that

    ● − □ = 8

    This can be rewritten as

    ● = 8 + □

    From row 4, we know that

    □ + □ + ● = 14

    Replacing the circle with 8 + □ gives us

    □ + □ + 8 + □ = 14

    □ + □ + □ = 6

    So, square = 2.

    From here, we can solve for the circle and then the triangle.

    NOTES

    This is a more advanced approach than the other methods because it involves moving a symbol from one side of an equal sign to the other. For students who are ready for it, this method is an excellent introduction to symbolic manipulation.

Explore More

10. If time permits, suggest that students try the Explore More. As explained in the directions on page "Make Your Own," students now create challenges for each other, choosing their own values for the circle, square, and triangle. Explain that students can enter any values they want, but the sketch will display those values rounded to the nearest whole number.

SUMMARIZE

Project the sketch on a large-screen display for viewing by the class. Expect to spend about 15 minutes.

11. Students should have their worksheets for this class discussion. Ask volunteers to explain how they determined the values of the three symbols, or unknowns. Students can include examples from the challenges they created on page "Make Your Own." To help students communicate clearly, have them record their approaches step-by-step for the class to follow. Working in the sketch, have the class solve a new problem using the approach described by the student.

12. Ask students to communicate in writing one of their solution strategies. ***Write an explanation for someone who has never used this sketch about how to use your method to find the values of the symbols.***

ANSWERS

2. Solution strategies will vary.

4. Problems and solutions will vary.

6. Problems and solutions will vary.

Circles, Squares, and Triangles

 Name: _____

In this activity you'll figure out the values of three unknowns.

EXPLORE

1. Open **Circles Squares Three Unknowns.gsp.**
   Go to page "Three Unknowns."

2. You will figure out the values of the circle, square, and triangle.
   All of the values are whole numbers between 0 and 10.

 Dragging a row of symbols across the line shows the sum.
Record the rows you choose to drag.
Explain how you find the values of the symbols.

_____

_____

_____

_____

_____

_____

_____

_____

_____

_____

3. To check your answer, press *Show Answers.*

4. Press *New Challenge.* Find the values of the symbols now.
   Record your work.

_____

_____

_____

_____

_____

_____

_____

_____

_____

_____

5. Press *Show Answers* to check your work.

EXPLORE MORE

6. Go to page "Make Your Own."
   Follow the directions on the page.

Odd Squares: Adding Consecutive Odd Numbers

INTRODUCE

Project the sketch for viewing by the class. Expect to spend about 10 minutes.

1. Before showing the sketch, write this problem on the board and discuss it.

$$1 + 3 + 5 + 7 + 9 + \cdots + 61 + 63 + 65 + 67$$

Can you describe the kind of numbers that are being added? Give students time to think, and then take responses. Students are likely to describe the numbers as consecutive odd numbers that begin with 1. Make sure all students understand the difference between *consecutive numbers* and *consecutive odd numbers.*

Also discuss the "..." notation. **This means that the pattern continues between 9 and 61. Because there are so many odd numbers in this sum, it's more convenient to just write "..." as shorthand notation.**

Adding all the odd numbers from 1 through 67 sounds like a lot of work! Today we're going to see whether we can use geometric and algebraic reasoning to find a simple and fast way to do this.

Introducing the Model

Record student responses throughout this discussion so students can reference them as they continue to work.

2. Distribute the worksheet. Open **Odd Squares.gsp** and go to page "1 + 3." Ask students to describe what is shown.

Have a volunteer press *Add 1 + 3.* **What can you say about the sum of 1 + 3 now?** Here are two student observations.

The groups come together to make a square that's 2 by 2.

2 times 2 is 4. So is 1 plus 3.

Explain that, in the table on the worksheet, students should record the dimensions of the large square, the sum of the odd numbers, and the number of consecutive odds being added together (the number of terms).

Rather than press *Add 1 + 3 + 5,* you can ask a volunteer to drag the L-shapes together to form the large square.

3. Go to page "1 + 3 + 5." **What do you predict will happen when I press the button this time?** Students are likely to predict that another large square will be formed, this time with dimensions 3 by 3. They may also offer the supporting reasoning that $3 \times 3 = 9$ and $1 + 3 + 5 = 9$. Press *Add 1 + 3 + 5* to verify. Have students fill out the second row of the table.

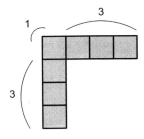

4. Repeat the above process on the remaining three pages of the sketch. Before showing each page, ask students to predict how the next odd number in the sequence will appear. For $1 + 3 + 5 + 7$, for example, the 7 will be represented as an L-shaped group of squares: three horizontal squares, three vertical squares, and one square at the corner. This pattern continues.

DEVELOP

Continue to project the sketch. Expect to spend about 20 minutes.

5. ***Do you see any patterns and relationships in the table and in the Sketchpad model?*** Provide time for students to work in pairs or small groups on this question, and then discuss. Elicit the ideas expressed in the sample student observations that follow. Make sure the last one is recorded where students can refer to it.

The sum of the consecutive odd numbers starting at 1 can be represented as a square.

The sum of the consecutive odd numbers starting at 1 can be represented as a square number $\left(4 = 2^2; 9 = 3^2; 16 = 4^2; 25 = 5^2; 36 = 6^2\right)$.

The number of odds being added is equal to the number of units for the side of the square.

Applying a Pattern

6. ***Let's see whether you can use that information to find the sum of a longer sequence. Can you find the sum of the first 20 consecutive odd numbers easily?*** Provide plenty of time for students, working in pairs or groups, to tackle this question. When students are ready, hold a discussion. A sample student explanation is this: *Because there are 20 odd numbers being added together, the large square will be 20 by 20. So, the sum of the first 20 consecutive odd numbers is 400.*

7. Return to the original problem. The challenge for students will be to determine how many odd numbers are being added. ***Work with a partner and use what you know now to help solve our original problem.***

$$1 + 3 + 5 + 7 + 9 + \cdots + 61 + 63 + 65 + 67$$

As a hint, you might ask, ***Suppose you were adding** all the numbers from 1 through 68, not just the odd numbers. How many numbers would you be adding?* Observe whether students understand how the answer to this question can help them. There are 68 numbers from 1 through 68, and half of those are odd numbers. So, there are 34 odd numbers from 1 through 68—and the same number of odd numbers from 1 through 67.

Have the class discuss their ways of solving the problem. [The sum is 34^2, or 1156.]

8. Continue by asking, ***How many odd numbers are in the sequence*** $1 + 3 + 5 + \cdots + 75 + 77?$ Observe whether students apply their reasoning from the previous problem to this new one. A sample student response is this: *Just like the last problem, we thought about the even number that is one more than 77. There are 78 numbers from 1 to 78. Half of those are odd numbers. So, there are 39 odd numbers from 1 to 78.*

Invite any students who have trouble understanding the equation to ask questions, so the class can help to clarify their understanding.

9. Write the following equation for students to consider. ***Does this equation represent your thinking?*** Ask several students to explain why this equation makes sense.

$$\frac{(77 + 1)}{2} = 39$$

Offer a few more sequences. Each time, ask students to write an equation like this one.

10. When most students understand the general pattern for determining how many odd numbers are in a given sequence, explain that they should use what they know to complete steps 2 and 3 on the worksheet then and check their answers with a partner.

SUMMARIZE

Continue to project the sketch. Expect to spend about 15 minutes.

11. Bring students together for a class discussion. They should have their worksheets with them. ***Can we write a rule that shows what you have been doing to find the sum?*** Record as a rule is formulated. Help the

class refine the wording. Here is a sample: *Take the last number in the sequence (the largest addend) and add 1. Divide that number by 2 to find out how many odd numbers there are. Square that number to find the sum of the sequence.*

12. ***Using this rule, let's write an expression for each row of the table.*** With students' help, record expressions for the sums of the consecutive odds.

$$\left(\frac{3+1}{2}\right)^2 \qquad \left(\frac{5+1}{2}\right)^2 \qquad \left(\frac{7+1}{2}\right)^2 \qquad \left(\frac{9+1}{2}\right)^2 \qquad \left(\frac{11+1}{2}\right)^2$$

13. ***Suppose we wanted to write an expression that shows how to find the sum of any sequence of consecutive odd numbers that begins at 1.*** Give students a minute to think about this proposition. ***Mathematicians have a way of doing this. They use letters to represent variables. Let's look at the expressions we just wrote. What varies in the expressions?*** [The number for the last addend in the sequence: 3, 5, 7, 9, 11] ***Let's use the letter n to stand for this last number in the sequence, the thing that varies.*** Record the following expression and ask students what they think it means.

$$1 + 3 + 5 + 7 + \cdots + n$$

Elicit the ideas that (1) the first four addends give enough information to tell us that this is a sequence of the consecutive odd numbers starting at 1, and (2) the n stands for the last addend in such a sequence. Students may suggest that n could be any odd number larger than 9. ***The variable n could be 11. It could be 27. It could be 999.***

14. ***For each row in the table, we wrote an expression for finding the sum. Can you write an expression like those expressions that tells how to find the sum for any sequence of consecutive odd numbers that begins at 1? Can a variable help you write the rule?*** Review the class's rule. Here is the sample from above: *Take the last number in the sequence (the largest addend) and add 1. Divide that number by 2 to find out how many odd numbers there are. Square that number.*

Have students work alone or in pairs to try to write an expression. Invite students to share theirs. If students have used the expressions for the table's rows and a variable for the last addend, their rules will look like the one shown here. (Using any letter for the variable is fine.

Some students may have used *a*, for *addend*, for example.) Have the class record this expression, or rule, on worksheet step 4.

$$\left(\frac{n+1}{2}\right)^2$$

15. ***Let's apply this rule. What is the sum of*** $1 + 3 + 5 + \cdots + 197 + 199$***?*** Discuss as a class. Invite a volunteer to record.

$$\left(\frac{199+1}{2}\right)^2 = 100^2 = 10{,}000$$

16. ***How would you explain to someone else the way this rule helps you find the sum of any sequence of consecutive odd numbers beginning at 1?*** You may wish to offer this question as a writing prompt and have students respond individually.

EXTEND

Working problems from multiple directions helps you and students assess their understanding.

Present the following problem: ***The sum of a sequence of consecutive odd numbers that starts with 1 is 81. What are the numbers in the sequence?*** Provide paper for students to work on and plenty of time to tackle this problem, which requires students to "undo" their rule, or work backward. When students are ready, invite them to discuss their reasoning. Here are sample student explanations.

If the sum is 81, then the large square formed by the L-shaped pieces must have 9 squares per side because 9 squared is 81. So, there are 9 numbers in the sequence. I wrote the first 9 consecutive odd numbers: 1, 3, 5, 7, 9, 11, 13, 15, 17.

When I look at our rule with n, I see that everything in the parentheses has to equal 9. So, some number divided by 2 has to equal 9. That number is 18. Then, some number plus 1 (n + 1) has to equal 18. That number is 17. The highest addend in the sequence is 17.

ANSWERS

1.

Addends	Large Square	Sum	Number of Terms
1 + 3	2 × 2	4	2
1 + 3 + 5	3 × 3	9	3
1 + 3 + 5 + 7	4 × 4	16	4
1 + 3 + 5 + 7 + 9	5 × 5	25	5
1 + 3 + 5 + 7 + 9 + 11	6 × 6	36	6

2. 169

3. 625

4. $\left(\dfrac{n+1}{2} \right)^2$

Odd Squares

Name:

In this activity you will use a pattern to help you find the sum of the consecutive odd whole numbers starting at 1.

1. Open **Odd Squares.gsp.** Use the pages of the sketch to fill in the table.

Addends	Large Square	Sum	Number of Terms
1 + 3			
1 + 3 + 5			
1 + 3 + 5 + 7			

2. What is the sum of 1 + 3 + 5 + ⋯ + 23 + 25?

3. What is the sum of 1 + 3 + 5 + ⋯ + 47 + 49?

4. What is the sum of $1 + 3 + 5 + 7 + \cdots + n$?

Signed Tiles: Evaluating Expressions

INTRODUCE

Project the sketch for viewing by the class. Expect to spend about 10 minutes.

1. Open **Signed Tiles.gsp** to page "Introduce." Enlarge the document window so it fills most of the screen.

2. Say, ***The tiles on this page represent an expression.*** Ask what the expression is. [$x + 2y + 4$] Manipulate the x slider and y slider so students see that the variables vary. ***What is the value of the expression when both $x = 2$ and $y = 2$?*** [10] ***When $x = 1$ and $y = 3$?*** [9]

3. ***We can check your answers by calculating the expression using Sketchpad's Calculator.*** Choose **Number | Calculate** and calculate the expression, clicking on the x- and y-values in the sketch to enter them in the expression. Once you have the expression, drag the sliders and say, ***To check the first answer, I drag the x slider to make x = 2, and I drag the y slider to make y = 2. We can see that the value of the expression for these x- and y-values is 10. Now if I change x to 1 and y to 3, the value of the expression is 9.***

4. Go to page "$2x + 3y + 3$" and page "New Expression" and say, ***As you work on this activity, the first couple of pages have tiles representing an expression you are to evaluate. Always evaluate the expression using paper and pencil first, then check your answer with Sketchpad's Calculator.***

5. Go to page "$x^2 + 2x + 3$" and say, ***Starting on page "x squared plus 2x plus 3," you will use custom tools to create your own expressions.*** Show the Custom Tools menu and the options available. Use the tools to create several example tiles, but not the actual expression on the page.

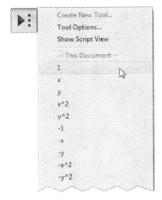

DEVELOP

Expect students at computers to spend about 25 minutes.

6. Assign student pairs to computers and tell them where to locate **Signed Tiles.gsp.** Distribute the worksheet and ask students to work on all steps. Encourage them to ask their classmates for help with Sketchpad.

7. Let pairs work at their own pace. As you circulate, here are some things to notice.

   • For students having trouble with the paper-and-pencil evaluations, suggest they write the expression replacing the variables with the

Exploring Expressions and Equations in Grades 6–8 with The Geometer's Sketchpad
© 2012 Key Curriculum Press

numbers before they multiply or combine terms. ***What is the value of this first term when $x = 2$?***

- On page "x^2 + 2x + 3" and beyond, students can choose to attach tiles to one another or not. If they make accidental tiles, remind them that they can choose **Edit | Undo**.

- On page "12," where $2x + y + 3$ is calculated, suggest that students leave x at one value and see what happens to the value of the expression as they move the y slider. ***What happens as y gets bigger? Why?*** Suggest they keep y constant and drag the x slider. ***How does dragging the x slider change the expression compared to dragging the y slider?***

- On page "13," where $3(x + 2) + y^2$ is calculated, students will have different ways of interpreting the parentheses and arranging the tiles. Make sure they have six units in their sketches. If they don't, ask them to explain how their sketch represents the expression and guide them with questions like ***Where are the three $(x + 2)$'s in your sketch?*** If their sketch is correct, you might ask, ***How does your arrangement represent the expression?***

- On page "14," where $4x - 2y + 4$ is calculated, make sure that students have used $-y$ tiles. Ask questions like ***What happens as y gets bigger? What happens as y gets smaller? Why? What does the $-2y$ part do when y is negative?***

8. Encourage students who finish quickly to try the Explore More suggestion. They can challenge each other with expressions of their own devising.

SUMMARIZE

Expect this part of the activity to take about 10 minutes.

9. ***What have you learned from working with the tiles?*** Elicit ideas such as these.

- The values of the variables determine the value of the expression. The value of an expression changes as the variables change.

- The custom tools in the Sketchpad algebra tiles can create expressions in x, y, x^2, y^2 and 1 and negative versions of these.

- The negative tile of a negative value represents a positive value.

- Expressions change more quickly when you drag the sliders when the slider variable is multiplied by a bigger number. (You could introduce the term *coefficient* here, if your curriculum uses the term.)

10. ***What other questions can you ask that might be explored?*** Encourage a variety such as these.

   Do some of these expressions have a smallest value or a largest value? Why?

   What kinds of expressions have a least value? What kinds have a largest value?

   Could we build an expression with x^3 or y^4 in it?

   Can expressions have more than two variables? Why are there only two variables in the custom tools?

   What are some of the expressions that can be made with the other custom tools?

ANSWERS

1. As the sliders are changed, tile lengths and numbers change.

2. $2x + 3y + 3 = 10$

3. Yes

4. a. 24   b. 26   c. 17

6. $3x + 6$

7. a. 18   b. 27   c. 3

10. a. 27   b. 66   c. 6

12. a. 14   b. 3   c. 9

13. a. 40   b. 16   c. 22

14. a. 6   b. −12   c. 24

15. Students' expressions will vary.

Signed Tiles

EXPLORE

1. Open the sketch **Signed Tiles.gsp.** The tiles on page "2x + 3y + 3" represent the expression 2x + 3y + 3. Drag the x and y sliders. What happens to the tiles and numbers in the sketch?

2. Use paper and pencil to evaluate the expression 2x + 3y + 3 for x = 2 and y = 1. What is the value?

3. To check your answer to question 2, choose **Number | Calculate.** Using ∗ for multiplication, enter the expression 2 ∗ x + 3 ∗ y + 3 in Sketchpad's Calculator. To enter the x and y in the expression, click on the x and the y in the sketch. Adjust the sliders to make x = 3 and y = 1. Does your answer to question 2 match Sketchpad's calculation?

4. Use paper and pencil to evaluate the expression 2x + 3y + 3 for these values.
   a. x = 3, y = 5   b. x = 1, y = 7   c. x = 4, y = 2

5. Adjust the sliders to check your answers to step 4.

6. Go to page "New Expression." What expression do the tiles on this page represent?

7. Use paper and pencil to evaluate your expression in step 6 for
   a. x = 4   b. x = 7   c. x = −1

8. Choose **Number | Calculate** and calculate the expression 3 ∗ x + 6. Drag the x slider to check your answers to step 7.

9. On page "x^2 + 2x + 3," you'll use custom tools to create tiles. Press and hold the **Custom** tool icon to see the menu of tools. Choose a tool and click on the vertical line to add a tile. Use the tools to create the expression $x^2 + 2x + 3$.

10. Use paper and pencil to evaluate the expression $x^2 + 2x + 3$ for

    a. $x = 4$ b. $x = 7$ c. $x = -3$

 11. Choose **Number | Calculate** and calculate the expression $x^2 + 2x + 3$.
    Drag the *x* slider to check your answers to step 10.

For questions 12–14, follow these steps.

   Create tiles to represent the expression.

   Use paper and pencil to evaluate the expression for each set of given values.

   Use Sketchpad's Calculator to calculate the expression.

   Drag sliders to check your answers to the question.

12. On page "12," create the expression $2x + y + 3$.

    a. $x = 3, y = 5$ b. $x = -2, y = 4$ c. $x = 4, y = -2$

13. On page "13," create the expression $3(x + 2) + y^2$.

    a. $x = 3, y = 5$ b. $x = -2, y = 4$ c. $x = 4, y = -2$

14. On page "14," create the expression $4x - 2y + 4$.

    a. $x = 3, y = 5$ b. $x = -2, y = 4$ c. $x = 4, y = -2$

EXPLORE MORE

15. On page "Explore More," use the tiles to create a complicated expression.
    Challenge yourself or a classmate to evaluate the expression for some given
    value or values. Check the answer using Sketchpad's Calculator.

Pool Border: Equivalent Expressions

 **ACTIVITY NOTES**

INTRODUCE

Project the sketch for viewing by the class. Expect to spend about 15 minutes.

1. Open **Pool Border.gsp** and go to page "Problem." Enlarge the document window so it fills most of the screen. Ask for a volunteer to read it aloud. Make sure everyone understands the problem.

2. Say, *Let's start with a 10-foot pool. What's a good problem-solving technique for getting started?* Most students will want to draw a diagram, if they haven't already done so. Have a student volunteer draw a diagram on the board.

3. *Can we solve the problem?* Some students may count tiles. Others may offer other answers, using multiplication and addition or subtraction. For each case, ask students to draw a diagram on the display with the regions shaded to illustrate their reasoning. Also work with the class to write each suggestion as an arithmetic expression involving the number 10. Encourage creative thinking. These are some of the many possibilities.

   - $10 + 10 + 10 + 10 + 4$
   - $4 \times 10 + 4$
   - $4 \times (10 + 1)$
   - $4 \times (10 + 2) - 4$
   - $2 \times (10 + (10 + 2))$
   - $12 \times 12 - 10 \times 10$

   If students omit parentheses and get an answer other than 44, take advantage of the situation to review standard order of operations. For example, if they write $4 \times 10 + 1$, they will get 41, because multiplication precedes addition; only with parentheses to make $4 \times (10 + 1)$ will the addition be done first.

4. *If the pool were a different size, we'd use a number other than 10. Would the expressions still give the same result? For example, if the side of the pool were any length x, would $4 \times x + 4$ equal $4 \times (x + 1)$?* Students will probably agree that the expressions would still be equal, because each expression represents the number of tiles in the same region. Some might even do a few calculations with various values of x.

5. *You can use Sketchpad to check for equivalence.* Go to page "Square." Mention that the displayed expression is an algebraic expression because it involves a variable. *Algebraic expressions are equivalent if they have the same values when the variable is replaced with any number.* Drag the x slider under the square to show that it controls the length of a side.

6. *Let's add another equivalent expression to the sketch. Which one shall we add from our list?* For example, if students pick $4 \times (10 + 2) - 4$, show how to choose **Number | Calculate** and enter into the Calculator $4 * (x + 2) - 4$. Then drag the slider for the square's side length to show that the values are the same. *What's the word again for expressions that have the same values for every number you substitute?* [Equivalent]

DEVELOP

Expect students at computers to work about 20 minutes.

7. Assign students to computers and tell them where to find **Pool Border.gsp.** Distribute the worksheet. Tell students to work through step 6 and do the Explore More if they have time.

8. As you circulate, help students enter new expressions as needed. Make sure they are drawing a diagram with shaded regions to illustrate each expression, but don't insist on any particular means of shading. Some of the diagrams may be ambiguous, but students should be able to explain their own drawings.

9. Have students put selected expressions and the accompanying diagrams on the board. Where diagrams don't represent the thinking clearly, ask for explanations. Students might display these expressions.

   • $4x + 4$

   • $4(x + 2) - 4$

   • $4(x + 1)$

   • $2(x + 2 + x)$

   • $2(2(x + 2) - 2)$

   • $(x + 2)^2 - x^2$

10. Ask students who worked on page "Rectangle" to share expressions for the number of border tiles in terms of side lengths m and n of the rectangle.

SUMMARIZE

Expect to spend about 10 minutes.

11. Reconvene the class. Have volunteers show their diagrams and explain their expressions.

12. *Let's use algebra to change expressions into other equivalent expressions. How might you change* $4(x + 1)$ *into the expression* $4x + 4$*?* Students might say, *Multiply through.* Review or introduce the term *distribute*.

13. Now work with an expression in which students combine like terms. *How might you change* $2(x + 2) + 2x$ *into the expression* $4x + 4$*?* Introduce or review the terminology *combining like terms*.

14. *What have you learned?* List all ideas students have. Bring out these objectives.

    • An algebraic expression involves variables.

    • If a number is written next to a variable, it means multiply.

    • An algebraic expression evaluates to a number when each variable is replaced by a number.

    • Algebraic expressions are equivalent if they have the same values when the variables are replaced by any number.

    • Distributing multiplication over addition gives equivalent algebraic expressions.

    • Combining like terms gives equivalent algebraic expressions.

15. *What other questions can you ask that might be explored?* Again, ideas will vary. Here are some interesting possibilities. You might suggest that students investigate these on their own.

    Why does distributing give equivalent expressions?

    Does addition distribute over multiplication?

    Do any other operations distribute over any operations?

 ACTIVITY NOTES

EXTEND

If students are quite comfortable with distributing and combining like terms, you might ask **Can you distribute in the expression $2(2(x + 2) - 2)$?** You might have half the class take on that expression and half work with $(x + 2)^2 - x^2$. Or, you might try reversing the distributivity to introduce simple factoring. **How might you start with $4x - 4$ and get to $4(x - 1)$?** Note the common factor of both terms and using it as a multiplier.

ANSWERS

Answers to steps 2, 4, and 5–7 will include expressions and diagrams for $4x + 4$, $4(x + 2) - 4$, $4(x + 1)$, $2(x + 2) + 2x$, $(x + 2)^2 - x^2$, and other expressions. For reference, go to page "Expressions."

Answer to step 8 will include expressions and diagrams for $2m + 2n + 4$, $2(m + 2) + 2n$, $2m + 2(n + 2)$, $2(m + 1) + 2(n + 1)$, and other expresssions.

Exploring Expressions and Equations in Grades 6–8 with The Geometer's Sketchpad

Pool Border

 Name:

In this activity you'll look for equivalent algebraic expressions for the number of tiles in a pool border.

EXPLORE

1. Open **Pool Border.gsp** and go to page "Square." One algebraic expression is already shown.

2. Look at the diagram of the pool and think of another way to combine the tiles. Choose **Number | Calculate** and enter an algebraic expression that represents that way of thinking. To enter *x* into the Calculator, click on the value in the sketch. What is your expression?

3. To change the size of the pool, drag the slider for *x*. Check that the value of your expression matches the value of the expression above. If not, change your expression until it does.

4. In the space below, draw a diagram that shows how you thought about the tiles. Write the expression under the diagram.

5. Repeat steps 2–4 for another expression. Draw a diagram and write the expression under the diagram.

6. Repeat steps 2–4 for another expression. Draw a diagram and write the expression under the diagram.

7. Repeat steps 2–4 for another expression. Draw a diagram and write the expression under the diagram.

8. Go to page "Rectangle." Now you can change both the width and the length of the pool. One algebraic expression for the number of border tiles is given. Add as many other equivalent expressions as you can think of. Check that they give the same values as the given expression when you drag the sliders. List your expressions.

EXPLORE MORE

9. Go to page "Expressions." Figure out the expression for each diagram, then press *Show* to check your answer. Finally, use algebra to change the expressions into the other expressions.

Exploring Expressions and Equations in Grades 6–8 with The Geometer's Sketchpad
© 2012 Key Curriculum Press

2

Operations

Broccoli and Brussels Sprouts: The Distributive Property

Students are introduced to the distributive property in the context of a garden area problem. They then generalize by calculating equivalent expressions with variables and model a number of distributive property equations using dynamic algebra tiles.

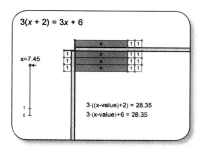

Algebars: Exploring Properties of Operations

Students compare algebraic expressions for equivalence using algebars, a Sketchpad model that represents the values of variables and expressions with bars. Changing the values of an algebraic expression is represented by varying the length of a bar. Students investigate the commutative, associative, and distributive properties.

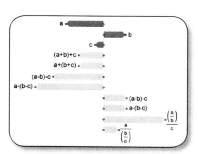

Powering Up: Multiplication and Exponents

Students calculate expressions for products of powers and dynamically test them for equivalence to other forms of the expression. They discover the law of exponents for multiplying powers with the same base.

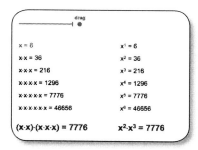

Powering Down: Division and Exponents

Students calculate expressions for quotients of powers and dynamically test them for equivalence to other forms of the expression. They discover the law of exponents for dividing powers of the same base.

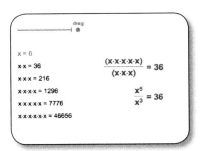

Power Strips: Laws of Exponents

Students use a Sketchpad model to explore repeated multiplication. They use custom tools to build expressions involving powers, and explore their behavior to find the multiplication, division, and power properties of exponents.

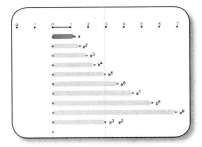

Exploring Expressions and Equations in Grades 6–8 with The Geometer's Sketchpad

Broccoli and Brussels Sprouts: The Distributive Property

For GSP5

INTRODUCE

Project the sketch for viewing by the class. Expect to spend about 10 minutes.

1. Open **Broccoli & Brussels Sprouts.gsp** and go to page "Diagram." Enlarge the sketch to fill the screen and say, *This diagram shows a garden that is divided into two parts, one for broccoli and one for Brussels sprouts. Each centimeter in the diagram represents one meter in the real garden. The head gardener and her assistant needed to find the total area of the garden. They calculated it in two different ways, but both got the same answer. What do you think the two ways were?* A student might answer, *One added 15 and 10 and then multiplied by 14, and the other multiplied 14 times 15 and 14 times 10 and then added.* Have students share, either verbally or by writing on the board, how the calculation methods give the same answer.

$$14 \cdot (15 + 10) = 14 \cdot 25 = 350$$

$$14 \cdot 15 + 14 \cdot 10 = 210 + 140 = 350$$

Expect students to write or say these calculations in different ways, but summarize by writing the equivalent calculations as an equation.

$$14 \cdot (15 + 10) = 14 \cdot 15 + 14 \cdot 10$$

2. Explain, *This is called the* **distributive property.** *Today you are going to use dynamic algebra tiles to model situations like these. Notice that the garden values can change.* Drag the red points to show that the measured values update to reflect the actual distances.

3. *On this page, you're going to make two calculations that both give an answer equal to the total area.* Demonstrate how to enter a measured value in a calculation. Choose **Number | Calculate** and click on a value in the sketch to enter it in the Calculator, but don't complete the expression—leave that for the students.

4. Go to page "Tiles." Explain, *In the rest of the activity, you'll be using dynamic algebra tiles to model distributive property expressions. On this page, the expressions are already modeled. The tiles on the outside of the corner piece represent the dimensions of the rectangle. The expression $3(x + 2)$ can be thought of as length multiplied by width. Your job will be to build and write the missing expression representing the area of the rectangle inside the frame.*

5. Go to page "Practice" and say, *On the rest of the pages, you'll use custom tools to construct the models yourself.* Demonstrate how to create the

arrangement shown on page "Tiles." You should also demonstrate the following mistake in some blank space on the page: *A common mistake is to try to line up unit tiles with the variable edge of an x tile. But notice that if x changes, the units don't line up anymore.*

You can line up unit tiles with the unit edges of the x tiles.

6. Go to page "GCF" and say, ***On this page you're given the rectangle and the area expression. You'll put tiles along the outside of the frame piece to represent the dimensions and write the expression length multiplied by width. On the rest of the pages, you'll be building the rectangles inside the frame and the dimensions along the outside of the frame.***

7. If you want students to save their work, demonstrate choosing **File | Save As,** and let them know how to name and where to save their files.

DEVELOP

Expect students at computers to spend about 30 minutes.

8. Distribute the worksheet and assign students to computers. Tell them where to locate **Broccoli & Brussels Sprouts.gsp.** Tell students to work through step 9 and do the Explore More if they have time. Encourage students to ask their neighbors for help if they are having difficulty with the construction.

9. Let pairs work at their own pace. As you circulate, here are some things to notice.

   • In worksheet steps 2 and 5, look for students who may need help entering an expression in the Calculator. A common mistake is to enter numbers or expressions instead of clicking on them in the sketch.

   • Watch for students who try to line up unit tiles with the variable dimension of an *x* tile. (See step 5 above.)

   • Remind students they can always undo if they make a mistake or accidentally create a stray tile.

- In parts of worksheet step 9, students will need to use the **x^2** tool to construct the rectangles in the frame.

- In worksheet steps 8 and 9, remind students to test their expressions for equivalence by calculating the two expressions on each page.

10. If students will save their work, remind them where to save it now.

SUMMARIZE

Project the sketch. Expect to spend about 5 minutes.

11. Gather the class. Display the sketch **Broccoli & Brussels Sprouts.gsp** again and go to page "Garden." Ask students what equal expressions they calculated. Write the equation $a(b + c) = ab + ac$, and identify this as the distributive property of multiplication over addition.

12. If a student did a Make Your Own problem, invite her to share it on the overhead display. Have her construct the model without revealing the expressions it represents; then invite the class to come up with the equation.

13. If your curriculum uses the vocabulary *factored form* and *expanded form*, this is a good opportunity to introduce it. Use a student example or an example from the worksheet. Use worksheet step 9e as an example. ***In the equation $x^2 + 7x = x(x + 7)$, the left side expression is called* expanded form *and the right side is called* factored form. *The factors are x and $x + 7$. In steps 9a through 9c, in what form were the given expressions?*** (Factored form) ***What about in steps 9d through 9f?*** (Expanded form)

ANSWERS

3. $a*(b + c) = a*b + a*c$

4. $3(x + 2) = 3x + 6$

8. $4(x + 1) = 4x + 4$

9. a. $3(2x + 1) = 6x + 3$
   b. $3x(x + 2) = 3x^2 + 6x$
   c. $x(2x + 5) = 2x^2 + 5x$
   d. $2x + 8 = 2(x + 4)$
   e. $x^2 + 7x = x(x + 7)$
   f. $3x^2 + 3x = 3x(x + 1)$

10. Possible expressions that can be modeled with rectangles are $3x(x + 1)$ and $x(3x + 3)$. Students might also observe that $3(x^2 + x)$ is another equivalent expression.

11. $a(b + c) = ab + ac$

Broccoli and Brussels Sprouts  Name:

The garden at right is divided into two parts, one for broccoli and one for Brussels sprouts. The head gardener and her assistant needed to know the total area of the garden. They each calculated it and got the same answer, but they did their calculations in two different ways.

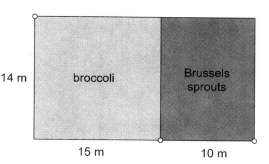

In this activity you'll model the distributive property—first with a dynamic garden, and then using dynamic algebra tiles.

EXPLORE

1. Open **Broccoli & Brussels Sprouts.gsp** and go to page "Diagram."

2. Choose **Number | Calculate** and enter an expression that is equal to the total area. Click on the measurements in the sketch to enter them into the Calculator.

3. Use the Calculator to enter a different expression that is equal to the total area. Drag points to confirm that both expressions are always equal to the total area. Write the equivalent expressions as an equation.

   _____ = _____

4. Go to page "Tiles." The tiles on the outside of the frame represent the dimensions of the rectangle inside the frame. Complete the equation by writing the expression represented by the tiles inside the frame.

   $3(x + 2) =$ _____

5. Choose **Number | Calculate** and enter the expression $3*(x + 2)$ into the Calculator. Click on the x-value on the slider to enter it in the calculation. Click **OK**.

6. Now enter the expression you wrote in step 4. Drag the x slider. Do the two expressions always remain equal?

7. Go to page "Practice." Practice using the custom tools to construct the arrangement on page "Tiles."

8. Go to page "GCF." The tiles inside the frame are represented by the expression $4x + 4$. Attach tiles along the outside of the frame to represent the dimensions of the rectangle. Complete the equation.

$$\underline{\hspace{3cm}} = 4x + 4$$

9. On pages "a" through "f," follow these steps.

- Use tiles to construct the dimensions along the outside of the frame and the rectangle inside the frame.

- Write the equation using the two representations of the area.

- Use Sketchpad's Calculator to calculate both expressions.

- Drag the x slider to check that the expression you wrote is always equivalent to the given expression.

    a. $3(2x + 1) = $ _____

    b. $3x(x + 2) = $ _____

    c. $x(2x + 5) = $ _____

    d. $2x + 8 = $ _____

    e. $x^2 + 7x = $ _____

    f. $3x^2 + 3x = $ _____

EXPLORE MORE

10. For the problem on page "f," there are two different-shaped rectangles you can make. Make the other possible rectangle and write the equation for this rectangle.

$$3x^2 + 3x = \underline{\hspace{3cm}}$$

11. Go to page "Garden" and represent the garden problem as an equation using only variables.

12. Go to page "Make Your Own." Use the tiles to create models of your own distributive property equation.

Exploring Expressions and Equations in Grades 6–8 with The Geometer's Sketchpad
© 2012 Key Curriculum Press

Algebars: Exploring Properties of Operations

INTRODUCE

Project the sketch for viewing by the class. Expect to spend about 10 minutes.

1. Open **Algebars.gsp.** Go to page "Intro." Enlarge the document window so it fills most of the screen.

2. Explain, *Today you're going to use Sketchpad to explore equivalent expressions. You'll use bars—called* **algebars**—*that represent variables and algebraic expressions.*

Review vocabulary as needed. A *variable* is a symbol, usually a letter, that represents a value that can change.

3. Press *Show Variables.* **The red bars are variables.** Drag point *a* and point *b* to the right and left to show how the lengths of the bars change. **What is the value of a? What is the value of b?** Have students practice reading the values of the variables as you drag the points to different locations. You might press *Show Vertical Lines* so that students can follow each gray vertical line up to find the value.

4. Drag point *a* to 0. **What happens to the red bar?** Students should observe that because the value is 0, the red bar disappears and becomes a point at 0.

An *algebraic expression* may contain variables, numbers, and operations.

5. Press *Show Algebars 1.* **The green algebars are expressions. What expressions do these green bars represent?** [$a + b$ and $b + a$] **What do you think will happen to the lengths of the green bars when a is 1 and b is 1?** Have students predict the lengths of each green bar. Then drag the points to check. Try other values for the variables, including negative values, and have students predict before you drag the points. Be sure students understand that the values of *a* and *b* are substituted into the expressions and affect their value. It's not necessary to point out that variables and numbers are also expressions.

6. **What did you notice about the lengths of the two green bars? Are these two expressions equivalent?** Drag the points again to change the values of *a* and *b*, if needed. Students should observe that for any values of *a* and *b*, the expressions are equivalent; the green bars are always the same length.

7. **How could you indicate that these two expressions are equivalent?** Students may make the following response: *Write it as an equation.* Write $a + b = b + a$ on the board after students suggest it.

8. Drag the points so that *a* and *b* are equal (a value close to 2 works well) and press *Show Algebars 2.* **What do these green bars represent?** [The expressions $a - b + 1$ and $b - a + 1$] **What do you think will happen to the lengths of the green bars when I drag points a and b?**

Have volunteers share their thoughts, and then drag the points to several different values. Be sure to include other examples for $a = b$. **What did you observe?** Students should respond that the lengths of the green bars are the same only when $a = b$. **Do you think that these expressions are equivalent? Can we use them to write an equation?** Students should realize that because the green bars are not the same length all the time, the expressions are not always equal, so an equation cannot be written.

9. *Now that you understand how algebars work, you can explore them on your own. You'll use them in this activity to explore some algebraic properties.*

10. If you want students to save their work in the Explore More, demonstrate choosing **File | Save As,** and let them know how to name and where to save their files.

DEVELOP

Expect students at computers to spend about 25 minutes.

11. Assign students to computers and tell them where to locate **Algebars.gsp.** Distribute the worksheet. Tell students to work through step 19 and do the Explore More if they have time.

12. Let pairs work at their own pace. As you circulate, here are some things to notice.

   • Encourage students to make predictions before dragging a and b. This will require them to think about how the variables and expressions are related.

   • In worksheet step 6, review the meaning of *commutative*. If an operation is commutative, it means that the operation can be performed in any order without affecting the outcome.

   • In worksheet step 7, as students are dragging, they should observe that the green bars for $a \div b$ and $b \div a$ will be equal lengths (that is, they both will be equal to 1) when $a = b$. **Does this mean that the expressions are equivalent? Explain.** This is a good check to be sure students understand that the green bars *always* need to be of equal length for the expressions to be equivalent.

   • In worksheet step 9, review the meaning of *associative*. If an operation is associative, changing the grouping of terms does not change the outcome.

- In worksheet steps 6 and 9, when the denominator is zero, the algebars disappear. Ask students why they think this happens with the model. Have them look at the expressions for division as a hint. It is not possible to divide by zero, so the algebars disappear whenever this happens.

- Review the order of operations as students work on the associative and distributive properties. ***What is the standard order of operations?*** [Operations within parentheses, exponents, multiplication and division from left to right, and then addition and subtraction from left to right] ***Why is it important for everyone to follow this order?*** [A standard rule ensures that everyone will get the same answer.]

- If students have time for the Explore More, they will learn how to construct algebars using **Custom** tools. After they build the algebars for the two expressions, students will explore whether they are equivalent. This is another investigation of the Distributive Property.

13. If students will save their work, remind them where to save it now.

SUMMARIZE

Project the sketch. Expect to spend about 10 minutes.

14. Gather the class. Students should have their worksheets with them. Open **Algebars.gsp** and go to page "Commutative." Begin the discussion by asking students to describe what equivalent expressions are. ***Today you used algebars to explore some equivalent expressions. What are equivalent expressions?*** Here are sample student responses.

    The green algebars, representing the expressions, stay the same length no matter where I drag the points representing the variables.

    They are expressions that are equal for any value of the variable.

15. Discuss worksheet steps 7, 10, 16, and 19 with the class. Have volunteers come up to the computer and prove which expressions are and are not equivalent. Summarize with the class by writing the properties on the board.

Property	Example
Commutative Property of Addition	$a + b = b + a$
Commutative Property of Multiplication	$a \cdot b = b \cdot a$
Associative Property of Addition	$(a + b) + c = a + (b + c)$
Associative Property of Multiplication	$(a \cdot b) \cdot c = a \cdot (b \cdot c)$
Distributive Property	$a(b + c) = ab + ac$

16. If time permits, discuss the Explore More. *Were the expressions equivalent? If so, what equation did you write? Can we add this equation to our chart?* Include $a(b - c) = ab - ac$ as another example of the Distributive Property. Explain that the Distributive Property works with subtraction as well because subtraction is the same as adding the opposite. Invite students to share other properties they may have explored.

EXTEND

1. Have students go to page "Make Your Own." Let students use the Sketchpad model to build algebars using the custom tools to discover whether these two expressions are equivalent: $2(a \div b)$ and $2a \div 2b$. Students will learn that the Distributive Property does not hold for multiplication over division. Have students share their thinking.

2. Have students explore which of the following expressions are equivalent: $a^c b^c$, $(a + b)^c$, and $(ab)^c$. They can build the expressions using the **(a^b)** tool. Students will learn that $a^c b^c = (a \cdot b)^c$ because exponents are distributive across multiplication, but not addition. Let volunteers share their results with the class.

3. *What other questions might you ask about equivalent expressions or properties of operations?* Encourage all inquiry. Here are some ideas students might suggest.

    If subtraction is the same as adding the opposite, why is subtraction not commutative?

    Is Corey's rule ever right? Are there any values of m and n for which $2 + mn = (2 + m)(2 + n)$?

    Why does multiplication distribute over addition?

    Why doesn't addition distribute over multiplication?

Exploring Expressions and Equations in Grades 6–8 with The Geometer's Sketchpad
© 2012 Key Curriculum Press

ANSWERS

3. The two expressions are equivalent because the green algebars always remain the same length. The equation is $a + b = b + a$.

5. The two expressions are not equivalent because the green algebars are not always the same length. They're the same only when $a = b$.

7. The addition algebars and the multiplication algebars are always the same length. The equations are $a + b = b + a$ and $ab = ba$.

8. Addition and multiplication are commutative; subtraction and division are not commutative.

10. The addition algebars and the multiplication algebars are always the same length. The equations are $(a + b) + c = a + (b + c)$ and $(a \cdot b) \cdot c = a \cdot (b \cdot c)$.

11. Addition and multiplication are associative; subtraction and division are not associative.

12. The expressions $2(c + 4)$ and $2c + 8$ are equivalent. As an equation, $2(c + 4) = 2c + 8$. You multiply 2 by each value in the parentheses and then add the results. This is an example of the distributive property of multiplication over addition.

13. Answers will vary. Some students may describe the behavior of the bars; others may give a counterexample; and others may give an algebraic argument in terms of the distributive property. Accept all reasonable answers. The main purpose is to get students to think about the question.

16. The two expressions are equivalent because the green algebars always stay the same length. The equation is $2(m + n) = 2m + 2n$. This is another example of the Distributive Property of Multiplication over Addition.

19. The two expressions, $2 + (m \cdot n)$ and $(2 + m) \cdot (2 + n)$, are not equivalent. The green algebars are not always the same length.

23. The two expressions are equivalent because the green algebars always stay the same length. The equation is $x(y - z) = xy - xz$. The Distributive Property works with subtraction as well because subtraction is the same as adding the opposite.

Algebars

In this activity you'll explore operations using algebars. Algebars are bars that represent algebraic quantities. Red bars are variables. Green bars are expressions. When two green bars are always equal in length, they represent equivalent expressions.

EXPLORE

1. Open **Algebars.gsp** and go to page "Intro." Press *Show Variables.* Two red algebars appear that represent the variables *a* and *b*. Drag points *a* and *b* left and right to change their values. You can press *Show Vertical Lines* to see their values more clearly.

2. Press *Show Algebars 1.* Two green algebars appear that represent the expressions $a + b$ and $b + a$. What do you think will happen to the lengths of the green algebars when you drag points *a* and *b*? Try it and check your prediction.

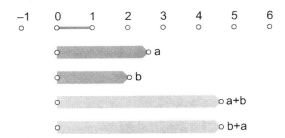

3. Are the two expressions equivalent? Explain. If so, write the result as an equation.

4. Press *Show Algebars 2.* Two green algebars appear that represent the expressions $a - b + 1$ and $b - a + 1$. What will happen to the green algebars when you drag points *a* and *b*? Try it and check your prediction.

5. Are the two expressions equivalent? Explain. If so, write the result as an equation.

Now you'll use algebars to explore some algebraic properties.

Commutative Property

6. Go to page "Commutative." This page shows four possible commutative properties.

Exploring Expressions and Equations in Grades 6–8 with The Geometer's Sketchpad
© 2012 Key Curriculum Press

Operation	Expressions
Addition	$a + b$ and $b + a$
Subtraction	$a - b$ and $b - a$
Multiplication	$a \cdot b$ and $b \cdot a$
Division	$a \div b$ and $b \div a$

Predict which green algebars will stay the same length when you drag points *a* and *b*. Then try it and observe what happens.

7. Which pairs of algebars are always the same length? Write an equation for each pair that matches.

8. Which of the four operations are commutative? Which are not?

Associative Property

9. Go to page "Associative." This page shows four possible associative properties.

Operation	Expressions
Addition	$(a + b) + c$ and $a + (b + c)$
Subtraction	$(a - b) - c$ and $a - (b - c)$
Multiplication	$(a \cdot b) \cdot c$ and $a \cdot (b \cdot c)$
Division	$(a \div b) \div c$ and $a \div (b \div c)$

Predict which green algebars will stay the same length when you drag points *a* and *b*. Then try it and observe what happens.

10. Which pairs of algebars are always the same length? Write an equation for each pair that matches.

11. Which of the four operations are associative? Which are not?

Distributive Property

12. Go to page "Distributive 1." Drag point c and observe what happens. Which two expressions are equivalent? Explain why. Write it as an equation.

13. Why is $2c + 4$ not equivalent to $2(c + 4)$?

14. Now go to page "Distributive 2." Sabrina says you can evaluate the expression $2(m + n)$ two different ways.

    $2(m + n)$: Add what's inside the parentheses first and then multiply by 2.

    $2m + 2n$: Multiply 2 by each value in the parentheses and then add the results.

15. Press *Show Sabrina's Algebars* to see how Sabrina tested her rule. The bottom two bars show the two methods.

16. Predict whether Sabrina is correct. Then drag points m and n and observe what happens. Are the two expressions equivalent? Explain. If so, write the result as an equation.

17. Corey says you can do something similar with the expression $2 + (m \cdot n)$.

    $2 + (m \cdot n)$: Multiply what's inside the parentheses first and then add 2.

    $(2 + m) \cdot (2 + n)$: Add 2 to each value in the parentheses and then multiply the results.

18. Press *Show Corey's Algebars* to see how Corey tested his rule. The bottom two bars show the two methods.

19. Predict whether Corey is correct. Then drag points *m* and *n* and observe what happens. Are the two expressions equivalent? Explain. If so, write the result as an equation.

EXPLORE MORE

Go to page "Explore More." Now you will build your own algebars to explore whether $x(y - z)$ and $xy - xz$ are equivalent expressions. You'll start by constructing the expression $(y - z)$.

20. Press and hold the **Custom** tool icon and choose the **(a−b)** tool. This tool requires you to click on five objects in this order:

   • an unused white point
   • the point at the tip of the *y* algebar
   • the caption of this point (*y*)
   • the point at the tip of the *z* algebar
   • the caption of this point (*z*)

21. To complete the expression $x(y - z)$, choose the **(a∗b)** tool and click on five objects in order, as in step 20, using the *x* algebar and the $(y - z)$ algebar that you just constructed.

22. Next you'll construct the expression $xy - xz$. Start by constructing xy and xz, using the **(a∗b)** tool on the *x*, *y*, and *z* algebars. Then construct $xy - xz$ by using the **(a−b)** tool on the $(x \cdot y)$ and $(x \cdot z)$ algebars you just constructed.

23. Now drag points *x*, *y*, and *z* and observe what happens. Are the two expressions equivalent? Explain. If so, write the result as an equation.

24. Use the custom tools to create algebars of your own to test for other equivalent expressions, including ones with exponents.

Powering Up: Multiplication and Exponents  ACTIVITY NOTES

INTRODUCE

Project the sketch for viewing by the class. Expect to spend about 5 minutes.

1. Introduce the objectives of the activity. *Today you'll review how to use exponents to represent repeated multiplication. Then you'll investigate the product of two powers that have the same base. I'll introduce the Sketchpad model, and then you'll work on your own.*

   Though some textbooks use the words *power* and *exponent* interchangeably, that usage can lead to confusion over "To multiply powers, add the exponents." To avoid any misunderstanding, keep the distinction clear between power and exponent as you talk with students. A *power* is a term that contains an *exponent*. For example, 10^3 is a power of 10; the exponent is 3. The expression $x + y^3$ includes a power of y. The expression $(x + y)^3$ represents the third power of $(x + y)$.

Students don't need to know how the measurement x is calculated from the slider for the purposes of this activity, so we refer to it simply as x.

2. Open **Powering Up.gsp.** Model worksheet steps 1–3. Show students how to use Sketchpad's Calculator. To create a calculation involving the x in the sketch, choose **Number | Calculate**, click on x in the sketch, use ∧ to create a power of x, and then click **OK**.

3. Change the value of x by dragging the slider to test for equivalence. *You can see that as I change the value of x, the values of the expressions $x \cdot x$ and x^2 are always equal, so they must be equivalent. Now you can try that on your own and practice using the Calculator to create powers for the rest of the repeated multiplication expressions. Then you can experiment with products.*

DEVELOP

Expect students at computers to spend 20 minutes.

4. Assign students to computers and tell them where to locate **Powering Up.gsp.**

5. Distribute the worksheet. Tell students to work through step 12 and do the Explore More if they have time. Encourage students to ask a neighbor for help if they have questions about using Sketchpad.

6. Let pairs work at their own pace. As you circulate, here are some things to notice.

   • In worksheet step 2, students may find that the sizes and styles of their calculations don't match the calculations already in the sketch. If they want their calculations to look similar to the ones in the sketch, show them how to choose **Display | Show Text Palette** and change the size and style of the calculation.

- In worksheet step 6, students don't need to create this calculation from scratch; if they try that, the Calculator will ignore the parentheses they enter. To show the parentheses, students should click on the existing expressions for $x \cdot x$ and $x \cdot x \cdot x$.

- In worksheet step 9, students should drag the slider to display the value of $x^2 \cdot x^3$ for several different values of x.

7. Encourage students to test their expressions frequently by dragging the slider. Ask questions such as, *Can you test whether or not your expressions are equal? What do you find when you test your expressions to see whether they're equal?* For students who are ahead of their peers, you might ask, *Are your expressions equal even when the base, x, is negative?*

8. Some students are likely to be able to fill out the table in worksheet step 10 without testing the expressions on Sketchpad. They won't be able to verify the equivalence dynamically, so ask them to write a description explaining why their calculation process works.

SUMMARIZE

Expect to spend about 5 minutes.

9. When most students have finished worksheet step 10, direct their attention away from the computers and ask them to share the expressions they wrote in the table. Record these expressions on the board as equations, for example, $x^1 \cdot x^3 = x^4$. Ask students to suggest a product that is not in the worksheet table and ask another student to provide the simplified form. *What is the product of $y^{20} \cdot y^4$?* $[y^{24}]$ *What is $2^5 \cdot 2^4$?* $[2^9]$ *What is $4^a \cdot 4^b$?* $[4^{a+b}]$

10. Ask students for their rule in worksheet step 11. Work toward a complete statement and summarize: *To multiply powers of the same base, add their exponents.* Discuss the symbolic representation in worksheet step 12. Write it on the board.

$$x^a \cdot x^b = x^{a+b}$$

Identify this as one of several *laws of exponents* and name it according to how your students' text names it, for example, the *product of powers property*. You might ask whether the rule could be stated using other letters, and let the class come up with some examples.

 ACTIVITY NOTES

11. Ask, *How would you describe the bases in the expression in step 8 on your worksheet?* [The expressions had the same base.] Explain that the base *x* could represent 2 or 3 or some other number.

12. Ask students whether they think the rule applies when bases are different. For example, *Can $x^3 \cdot y^2$ be simplified?* Students who had time to work on the Explore More may know that it cannot. Others may start with misconceptions, such as $x^3 \cdot y^2 = (xy)^5$. But if you ask them to defend their theories and write out the expressions as repeated multiplication, they will probably conclude that the law applies only to like bases and that $x^3 \cdot y^2$ cannot be simplified.

13. If time permits, give students some guided practice finding products of powers, perhaps including some monomials with numerical coefficients and/or more than one base.

ANSWERS

5. $x \cdot x = x^2$; $x \cdot x \cdot x = x^3$; $x \cdot x \cdot x \cdot x = x^4$; $x \cdot x \cdot x \cdot x \cdot x = x^5$; $x \cdot x \cdot x \cdot x \cdot x \cdot x = x^6$; $x \cdot x \cdot x \cdot x \cdot x \cdot x \cdot x = x^7$

7. $(x \cdot x) \cdot (x \cdot x \cdot x) = x^2 \cdot x^3$

8. $(x \cdot x)(x \cdot x \cdot x) = x^2 \cdot x^3 = x^5$

10.

Product	Product of Powers $x^? \cdot x^?$	Simplified Power $x^?$
$(x) \cdot (x \cdot x \cdot x)$	$x^1 \cdot x^3$	x^4
$(x \cdot x) \cdot (x \cdot x \cdot x \cdot x)$	$x^2 \cdot x^4$	x^6
$(x \cdot x) \cdot (x \cdot x)$	$x^2 \cdot x^2$	x^4
$(x \cdot x \cdot x \cdot x) \cdot (x \cdot x \cdot x)$	$x^4 \cdot x^3$	x^7
$(x \cdot x \cdot x \cdot x) \cdot (x \cdot x \cdot x \cdot x)$	$x^4 \cdot x^4$	x^8
$(x \cdot x) \cdot (x \cdot x) \cdot (x \cdot x \cdot x)$	$x^2 \cdot x^2 \cdot x^3$	x^7

11. When you multiply powers with like bases, you add exponents.

12. $x^a \cdot x^b = x^{a+b}$

13. Students who try the Explore More should discover that products of powers with different bases can't be multiplied by adding the exponents.

Exploring Expressions and Equations in Grades 6–8 with The Geometer's Sketchpad
© 2012 Key Curriculum Press

Powering Up

 Name:

In this activity you'll experiment with multiplying powers with the same base. You'll discover a rule for writing products of powers.

EXPLORE

1. Open **Powering Up.gsp** and go to page "Multiplying Powers." You'll see the equation $x = 2$ and equations containing two factors of x, three factors of x, and other products of repeated multiplication.

 2. One factor of x can be written with an exponent, x^1, as shown in the sketch. Choose **Number | Calculate**. In the Calculator, create an expression that uses an exponent and is equivalent to $x \cdot x$.

   To enter x into the Calculator, click on the equation $x = 2$ in the sketch. To raise it to a power, click ^ on the Calculator keypad.

   When you're finished, click **OK**. If necessary, drag the calculation to make it line up nicely under $x^1 = 2$.

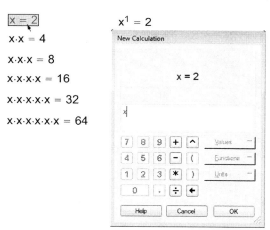

3. To check whether the new expression is equivalent to $x \cdot x$, change the value of x by dragging the slider, and look at the values of both expressions.

4. Is the expression you created equivalent to $x \cdot x$?

5. Repeat step 2 to create expressions with exponents that are equivalent to each of the other products of repeated multiplication. Check for equivalence by changing the value of x.

6. Now you'll experiment with products of powers. Choose **Number | Calculate** and click on $x \cdot x$, then *, and then $x \cdot x \cdot x$ to create the expression $(x \cdot x)(x \cdot x \cdot x)$. Click **OK.**

7. To create an expression that is equal to the expression in step 6, choose **Number | Calculate** and enter the product of two powers of x.

   Write this equation with the correct exponents.
   $$(x \cdot x) \cdot (x \cdot x \cdot x) = x^? \cdot x^?$$

8. What expression with a single exponent is the product in step 6 equal to? Write this equation with the correct exponents.
   $$(x \cdot x) \cdot (x \cdot x \cdot x) = x^? \cdot x^? = x^?$$

9. Check that your expressions in step 8 are equal by changing the value of x.

10. Repeat steps 6–9 for each of the products in the table. Drag the slider to test that both expressions are always equal. Record your expressions in the table.

Product	Product of Powers $x^? \cdot x^?$	Simplified Power $x^?$
$(x) \cdot (x \cdot x \cdot x)$		
$(x \cdot x) \cdot (x \cdot x \cdot x \cdot x)$		
$(x \cdot x) \cdot (x \cdot x)$		
$(x \cdot x \cdot x \cdot x) \cdot (x \cdot x \cdot x)$		
$(x \cdot x \cdot x \cdot x) \cdot (x \cdot x \cdot x \cdot x)$		
$(x \cdot x) \cdot (x \cdot x) \cdot (x \cdot x \cdot x)$		

11. Look for patterns in your table. Describe how to multiply powers with the same base.

12. Replace the question mark to generalize your observation as a rule.
    $$x^a \cdot x^b = ?$$

Exploring Expressions and Equations in Grades 6–8 with The Geometer's Sketchpad
© 2012 Key Curriculum Press

EXPLORE MORE

13. Choose **Number | New Parameter** and name the new variable y. Make its value different from x. Calculate a power of y.

    Experiment with products of powers of x and y. Is it possible to simplify a product of two powers such as $x^2 \cdot y^3$? Can you create an equivalent expression with just one exponent? Explain.

Powering Down: Division and Exponents

 **ACTIVITY NOTES**

INTRODUCE

Project the sketch for viewing by the class. Expect to spend about 5 minutes.

1. Students should have completed the activity Powering Up. Explain, *Today you'll review a simple but important rule about dividing powers. Then you'll experiment with dividing two powers of the same base. I'll show you how to divide using Sketchpad's Calculator, and then you'll work on your own.*

   Though some textbooks use the words *exponent* and *power* interchangeably, that usage can lead to confusion over "To divide powers, subtract the exponents." Make the distinction in your own use of the words: A *power* is a term that contains an *exponent*. For example, 10^3 is a power of 10, and the exponent is 3. The expression $x + y^3$ includes a power of y. The expression $(x + y)^3$ represents the third power of $(x + y)$.

Students don't need to know how the measurement x is calculated from the slider for the purposes of this activity, so we refer to it simply as x.

2. Open **Powering Down.gsp.** Model worksheet steps 1–3. Show students how to use Sketchpad's Calculator. To create a calculation involving the x in the sketch, such as $\frac{x}{x}$, choose **Number | Calculate**, click on x in the sketch, then ÷ on the Calculator keypad, then x again, and then click **OK.** Pressing *Show Powers* displays the expressions using exponents.

3. *What do I get if I divide x by x? Does it matter what the value of x is?* Demonstrate how to change the value of x by dragging the slider. *What number should I substitute for x?* Try several values students suggest. If no student suggests 0, you might suggest it to show that $\frac{0}{0}$ is undefined. You might ask, *Why is $\frac{0}{0}$ undefined?* [Because any number times 0 is 0, there is no way to determine the answer: $0 \cdot 5 = 0$; $0 \cdot 154 = 0$; $0 \cdot n = 0$, so $\frac{0}{0}$ is 5 or 154 or any number.]

4. *As long as x is not 0, when I divide it by itself, I get 1. Can you think of an example in arithmetic where you use this idea?* If necessary, follow up with, *How do we use this idea when we reduce fractions?*

5. *You will now use the idea of dividing out common factors as you simplify expressions that involve quotients of powers.*

DEVELOP

Expect students at computers to spend 20 minutes.

6. Assign students to computers and tell them where to locate **Powering Down.gsp.**

7. Distribute the worksheet. Tell students to work through step 12 of the worksheet and do the Explore More if they have time. Encourage

Exploring Expressions and Equations in Grades 6–8 with The Geometer's Sketchpad
© 2012 Key Curriculum Press

 ACTIVITY NOTES

students to ask a neighbor for help if they have questions about using Sketchpad.

8. Let pairs work at their own pace. As you circulate, here are some things to notice.

   - As students work on worksheet step 4, ask, **When have you used this law of arithmetic?** [Reducing fractions]

   - In worksheet step 5, students can create the calculation from scratch or they can use the existing expressions, for $x \cdot x \cdot x \cdot x \cdot x$ and $x \cdot x \cdot x$. If they use the existing expressions, the numerator and denominator will appear with parentheses; if they create a calculation from scratch, there will be no parentheses.

   - In worksheet step 8, students should calculate $\frac{x^5}{x^3}$ and compare it to x^2.

   - For the expression $\frac{x \cdot x \cdot x \cdot x \cdot x}{x}$ in worksheet step 9, ask, **How can the denominator, x, be written with an exponent?** [x^1]

9. Encourage students to test their expressions by changing the value of x. **What do you find when you test your expressions to see whether they're equivalent?**

10. Some students may be able to fill out the table in worksheet step 9 without testing the expressions on Sketchpad. Ask them to explain why their methods work. To add more challenges, you might ask, **Are your expressions equivalent even when the base, x, is negative?**

SUMMARIZE

Expect to spend
5 minutes.

11. When students have finished worksheet step 10, direct their attention away from the computers and write $\frac{x \cdot x \cdot x \cdot x \cdot x}{x \cdot x \cdot x}$ on the board. Ask them to explain why $\frac{x^5}{x^3} = x^2$. Be sure the discussion includes that $\frac{x}{x}$ equals 1, so each pair of factors of x in the numerator and denominator divide to make 1. (That is, the x's "cancel.") For example, in worksheet step 7, two factors of x remain in the numerator.

12. Have students share the expressions they wrote in the table in worksheet step 9. Record these expressions on the board as equations: $\frac{x^5}{x^2} = x^3$, for example.

13. Ask what students gave for their rule in worksheet step 11. Write it on the board.

$$\frac{x^a}{x^b} = x^{a-b}$$

Identify this as one of several *laws of exponents* and name it according to its name in your students' text, for example, the *quotient of powers property*. Clearly state the rule in words. **To divide powers of the same base, subtract the exponents.** Include the phrase *of the same base*.

14. Point out that in the investigation, all the expressions had the same base, *x*. Ask students, **Do you think the rule applies when bases are different? For example, can you simplify $\frac{x^3}{y^2}$?** Students may start with misconceptions, but if you ask them to defend their theories, they should conclude that no common factors will divide out, and the law only applies to like bases.

15. If time permits, give students some guided practice dividing powers, perhaps including monomials with numerical coefficients or expressions with more than one base.

16. If negative exponents are part of your curriculum, discuss Explore More.

ANSWERS

4. Any number divided by itself is equal to 1.

6. $\dfrac{x \cdot x \cdot x \cdot x \cdot x}{x \cdot x \cdot x} = x \cdot x = x^2$

7. The three factors of *x* in the numerator divided by three factors of *x* in the denominator equal 1. Two factors of *x* remain in the numerator, so the quotient equals $x \cdot x$.

8. $\dfrac{x \cdot x \cdot x \cdot x \cdot x}{x \cdot x \cdot x} = \dfrac{x^5}{x^3} = x^2$

 ACTIVITY NOTES

9.

Quotient Using Repeated Factors	Equivalent Quotient Using Exponents $\dfrac{x^?}{x^?}$	Equivalent Power $x^?$
$\dfrac{x \cdot x \cdot x \cdot x \cdot x}{x \cdot x}$	$\dfrac{x^5}{x^2}$	x^3
$\dfrac{x \cdot x \cdot x \cdot x}{x \cdot x \cdot x}$	$\dfrac{x^4}{x^3}$	x^1 or x
$\dfrac{x \cdot x \cdot x \cdot x \cdot x}{x}$	$\dfrac{x^5}{x^1}$	x^4
$\dfrac{x \cdot x \cdot x \cdot x \cdot x \cdot x}{x \cdot x}$	$\dfrac{x^6}{x^2}$	x^4
$\dfrac{x \cdot x \cdot x}{x \cdot x \cdot x}$	$\dfrac{x^3}{x^3}$	x^0 or 1

10. The exponent in the equivalent power is the difference between the exponent in the numerator and the exponent in the denominator.

11. $\dfrac{x^a}{x^b} = x^{a-b}$

12. $x^0 = 1$

13. They should conclude that $x^{-n} = \dfrac{1}{x^n}$.

Powering Down

In this activity you'll experiment with dividing powers with the same base. You'll discover a rule for writing quotients of powers.

EXPLORE

1. Open **Powering Down.gsp** and go to page "Dividing Powers." You'll see the equation $x = 4$ and equations containing two factors of x, three factors of x, and other products of repeated multiplication. Each repeated multiplication of x can be written as a power of x.

 2. Now you will create the calculation $\frac{x}{x}$. Choose **Number | Calculate**. In the Calculator, enter x by clicking on the equation $x = 4$ in the sketch. Then click ÷ on the Calculator keypad, and then click on x again.

   When you're finished, click **OK**.

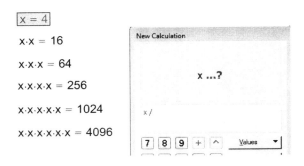

3. Change the value of x by dragging the slider.

4. What basic law of arithmetic does this calculation demonstrate?

5. Now, you'll experiment with dividing powers. Choose **Number | Calculate** and click on $x \cdot x \cdot x \cdot x \cdot x$ in the sketch, ÷ on the keypad, $x \cdot x \cdot x$ in the sketch, and then **OK** to create the expression

$$\frac{x \cdot x \cdot x \cdot x \cdot x}{x \cdot x \cdot x}$$

6. The calculation should be equal to one of the calculations already in your sketch. Which one? Drag the slider to confirm that these expressions are always equal.

7. How does your answer to step 4 help explain why $\frac{x \cdot x \cdot x \cdot x \cdot x}{x \cdot x \cdot x} = x \cdot x$?

Exploring Expressions and Equations in Grades 6–8 with The Geometer's Sketchpad
© 2012 Key Curriculum Press

8. Choose **Number | Calculate.** Create another expression equivalent to $\frac{x \cdot x \cdot x \cdot x \cdot x}{x \cdot x \cdot x}$ in the form of $\frac{x^?}{x^?}$. Use the x in the sketch and the $\wedge$ on the Calculator keypad to enter powers. As you finish each expression, click **OK.**

    Write this equation with the correct exponents.

    $$\frac{x \cdot x \cdot x \cdot x \cdot x}{x \cdot x \cdot x} = \frac{x^?}{x^?} = x^?$$

9. Repeat steps 5 and 8 for each of the quotients in the table. Drag the slider to test that both expressions are always equal. Record your expressions in the table.

Quotient Using Repeated Factors	Equivalent Quotient Using Exponents $\dfrac{x^?}{x^?}$	Equivalent Power $x^?$
$\dfrac{x \cdot x \cdot x \cdot x \cdot x}{x \cdot x}$		
$\dfrac{x \cdot x \cdot x \cdot x}{x \cdot x \cdot x}$		
$\dfrac{x \cdot x \cdot x \cdot x \cdot x}{x}$		
$\dfrac{x \cdot x \cdot x \cdot x \cdot x \cdot x}{x \cdot x}$		
$\dfrac{x \cdot x \cdot x}{x \cdot x \cdot x}$		

10. Look for patterns in your table. Describe how to divide powers with the same base.

11. Replace the question mark to generalize your observation as a rule.

    $$\frac{x^a}{x^b} = ?$$

12. The last row of your table demonstrates a special case of the rule that you might find surprising. Complete this equation.

    $$x^0 = \underline{\quad}$$

Powering Down

continued

EXPLORE MORE

13. In all the quotients in step 9, the power in the numerator has a larger exponent than the power in the denominator. What if the power in the denominator had the larger exponent? Experiment calculating expressions like $\frac{x^3}{x^6}$. Is the generalization you made in step 11 still true? Experiment calculating powers with negative exponents. Complete this equation.

    $$x^{-n} = \underline{\hphantom{mm}}$$

Exploring Expressions and Equations in Grades 6–8 with The Geometer's Sketchpad
© 2012 Key Curriculum Press

Power Strips: Laws of Exponents

INTRODUCE

Project the sketch for viewing by the class. Expect to spend about 10 minutes.

1. Open **Power Strips.gsp** and go to page "Patterns." Enlarge the document window so it fills most of the screen.

2. Explain, *Today you're going to use Sketchpad to explore operations involving exponents. You'll look for patterns and make conjectures.*

3. Slowly drag the red point *x* left and right on the sketch. **What do you observe? What do you think the points represent?** Let students share their ideas; don't give specific feedback at this time. These questions encourage students to observe closely and think about the patterns they see.

4. Press *Show Expressions*. Help students recognize that the points represent various powers of *x*. The vertical position corresponds to the power (exponent) and the horizontal position corresponds to the value of *x*. Tell students they will use this model to explore operations with exponents.

5. Continue dragging the red point *x*. **How many different positions make all the points line up? What values of x do you think these positions indicate?** [The points all line up when $x = 0$ and $x = 1$.] Let students conjecture, then press *Show Numbers*. Have students explain why the points should line up at these particular values.

6. **These expressions can be abbreviated using exponents. Can you explain what an exponent means?** Work with the class to come up with a definition of *exponent* and write it on the board. Here is a sample definition: *An exponent is a mathematical notation that means a number is multiplied by itself repeatedly. The base is the number to be multiplied, and the exponent is how many times the base is a factor. The number expressed as a base raised to an exponent is called a power.* Show an example, such as 5^3. Make sure students understand that it represents $5 \cdot 5 \cdot 5$, that 5 is the base, and that 3 is the exponent. The expression 5^3 can be read "the third power of 5" or "5 to the third power." The latter uses the word *power* to mean *exponent*, which may be confusing to students.

7. Go to page "Multiplication." Model worksheet step 9, *Now I'll show you how to use one of the custom tools, the a*b tool. After you choose this tool, you must click on five objects in this order: (1) an unused point, (2) the point of the first strip, (3) the label of this point, (4) the point of*

the second strip, and (5) the label of this point. Notice that the tool name includes these five steps. Emphasize that an "unused point" is one of the small circles that is not already the left endpoint of a strip. Make sure you have practiced using the tool beforehand. The other custom tools work in the same way.

8. If you want students to save their work, demonstrate choosing **File | Save As,** and let them know how to name and where to save their files.

DEVELOP

Expect students at computers to spend about 25 minutes.

9. Assign students to computers and tell them where to locate **Power Strips.gsp.** Distribute the worksheet. Tell students to work through step 22 and do the Explore More if they have time. Encourage students to ask their neighbors for help if they are having difficulty with Sketchpad.

10. Let pairs work at their own pace. As you circulate, here are some things to notice.

- In worksheet steps 5 and 6, be sure students notice that the differences in the lengths of the strips are not constant. As x moves from 1 to the right, the lengths of the strips are increasing. As x moves from 1 to 0, the lengths of the strips decrease. *Are the differences in the lengths of the strips constant from one strip to the next?*

- In worksheet step 7, students should notice that the strips alternate, pointing right and then left. *What is the sign of the product when you multiply two negative numbers? When you multiply three negative numbers?*

- Before students go to worksheet step 8, have students tell you how to write the expression $x \cdot x \cdot x \cdot x$ using exponents. [x^4] This will help them connect page "Multiplication" with page "Patterns." The model is the same, but the "Multiplication" page uses exponents.

- In worksheet steps 9–22, students will use custom tools to multiply and divide powers with the same base and find the value of a power raised to a power. Have students help each other use the tools if questions arise.

- In worksheet step 12, encourage students to drag x back and forth to check that the x^5 strip is the only strip that is the same length. *When you test, be sure to drag x left and right to try different values.*

Ask students to write out x^3, x^2, and x^5 using repeated multiplication to see why the strip for x^5 matches.

- In worksheet steps 16 and 20, have students use repeated multiplication to see why these problems work as well. ***Can you rewrite these expressions without using exponents?***

- In worksheet steps 14, 18, and 22, students will try to make conjectures after observing the simplification of two sample expressions. Encourage students to construct as many expressions as needed to make and test their conjectures. ***If you don't see a pattern yet, make another expression. Continue until you can make a conjecture. Then test it.***

- If students have time for the Explore More, they will investigate what happens when multiplying powers with different bases. Students will discover that there is no general pattern, so no rule can be made about simplifying a problem like $x^a \cdot y^b$.

11. If students will save their work, remind them where to save it now.

SUMMARIZE

Project the sketch. Expect to spend about 10 minutes.

12. Gather the class. Open **Power Strips.gsp.** Students should have their worksheets with them. Begin the discussion by asking students to share their conjectures. ***What conjectures did you make today?*** Volunteers may wish to test their conjectures for the class by constructing a problem using the appropriate page on the sketch.

13. When the class agrees on a conjecture, write it on the board using words and symbols. You may need to help students observe that the bases were the same in the problems. ***What do you notice about the bases in all the problems?*** Here is a sample summary of the properties with exponents covered in this activity.

Words	Symbols
When multiplying two powers with the same base, keep the base and add the exponents.	$x^a \cdot x^b = x^{a+b}$
When dividing two powers with the same base, keep the base and subtract the exponents.	$\dfrac{x^a}{x^b} = x^{a-b}$
When raising a power to a power, keep the base and multiply the exponents.	$(x^a)^b = x^{ab}$

14. If time permits, discuss the Explore More. *What did you notice when the bases were different? Could you still use your conjecture from worksheet step 14?* It is important that students understand that the bases need to be the same for properties to hold true. There is no rule for simplifying problems with different bases.

15. You may wish to have students respond individually in writing to this prompt. *Are $(x^a)^b$ and $x^a \cdot x^b$ the same value? Explain.*
[No, $(x^a)^b = x^{ab}$ and $x^a \cdot x^b = x^{a+b}$.]

EXTEND

What questions occurred to you about operations with exponents?
Encourage curiosity. Here are some sample student queries.

Why do the rules hold?

What happens when the exponents are negative? Do the properties still work?

Is $(x^a)^b$ the same as $(x^b)^a$?

Can exponents ever be fractional, between integers?

When an exponent is 0, the expression simplifies to 1. Does this affect the properties at all?

What happens when you have a more complicated expression like $(x^2 y^3)^2$?

How do you simplify an expression like $\left(\frac{3x^2}{y^2}\right)^3$?

ANSWERS

3. Answers will vary. The points represent various powers of *x*, with the vertical position corresponding to the power and the horizontal position corresponding to the value of *x*.

4. There are two positions of *x* at which all the points line up: $x = 1$ and $x = 0$. Repeatedly multiplying either 1 or 0 by itself continues to give the same result.

5. When $x > 1$, the points move increasingly rightward as you go down the screen, showing that the value of x^n increases more and more quickly for greater and greater values of *n*.

6. When $0 < x < 1$, the points move toward the left as they go down, approaching a straight line at $x = 0$. This pattern makes sense because multiplying by a value less than 1 always gives a result that is closer to 0 than the number you started with.

7. When $x < 0$, the strips alternate between the left and right sides because multiplying a number by a negative value always gives a result with a sign opposite to the sign of the original number.

10. The $x^3 \cdot x^2$ strip is the same length as the x^5 strip.

12. The $x^3 \cdot x^2$ strip is the same length as the x^5 strip no matter how you drag the value of x. This makes sense because you've multiplied three x's by two more x's, so that there are now five x's multiplied together.

13. The strip for $x^4 \cdot x^3$ is the same length as the x^7 strip.

14. Conjecture: The strip for $x^a \cdot x^b$ is the same length as the x^{a+b} strip. When multiplying powers with the same base, keep the base and add the exponents. Students will construct different products to test this conjecture.

16. When you use the **a/b** tool to create a strip for $\frac{x^7}{x^3}$, the resulting strip is the same length as the x^4 strip.

17. The resulting strip is the same length as the x^6 strip.

18. Conjecture: The strip for $\frac{x^a}{x^b}$ is the same length as the x^{a-b} strip. When dividing powers with the same base, keep the base and subtract the exponents. Students will construct different quotients to test this conjecture.

20. When you use the **(a)^b** tool to create a strip for $(x^4)^2$, the result matches x^8.

21. The resulting strip is the same length as the x^9 strip.

22. Conjecture: The strip for $(x^a)^b$ is the same length as an x^{ab} strip. When raising a power to a power, keep the base and multiply the exponents. Students will construct different powers of powers to test this conjecture.

23. If the bases are not the same, there is no general rule you can use to simplify a problem like $x^a \cdot y^b$. The special case of $x^a \cdot y^a = (xy)^a$ does not generalize to cases in which $a \neq b$, as testing with many values of x, y, a, and b will confirm.

Power Strips

In this activity you'll use a visual model to explore operations involving exponents.

EXPLORE

1. Open **Power Strips.gsp.** Go to page "Patterns."

2. Drag the red point *x* left and right. Observe what happens to the green points.

3. Press *Show Expressions.* Explain what the points represent.

4. How many positions can you find that make all the points line up? What values of *x* do you think these positions indicate? Press *Show Numbers.* Explain why the points line up for these values.

x

$x \cdot x$

$x \cdot x \cdot x$

$x \cdot x \cdot x \cdot x$

$x \cdot x \cdot x \cdot x \cdot x$

$x \cdot x \cdot x \cdot x \cdot x \cdot x$

$x \cdot x \cdot x \cdot x \cdot x \cdot x \cdot x$

$x \cdot x \cdot x \cdot x \cdot x \cdot x \cdot x \cdot x$

$x \cdot x \cdot x \cdot x \cdot x \cdot x \cdot x \cdot x \cdot x$

$x \cdot x \cdot x \cdot x \cdot x \cdot x \cdot x \cdot x \cdot x \cdot x$

5. What pattern do the points make when $x > 1$? Why does this make sense?

6. What pattern appears when $0 < x < 1$? Explain this pattern.

7. What pattern appears when $x < 0$? Explain this pattern.

8. Go to page "Multiplication." This page uses exponents to write the multiplication problems more simply. Drag point *x.* You should observe the same behaviors as on page "Patterns."

Exploring Expressions and Equations in Grades 6–8 with The Geometer's Sketchpad
© 2012 Key Curriculum Press

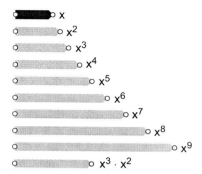

9. Choose the **a∗b** custom tool from the Custom Tools menu. Use the **a∗b** tool to multiply $x^3 \cdot x^2$ by clicking on these five points in order:

   - The first unused point below all the strips
   - The point at the tip of the x^3 strip
   - The x^3 label
   - The point at the tip of the x^2 strip
   - The x^2 label

10. Is the $x^3 \cdot x^2$ strip the same length as any of the existing strips?

11. Test your answer by clicking the **Indicator** custom tool on the tip of the $x^3 \cdot x^2$ strip.

12. Drag x and describe your observations.

13. Multiply $x^4 \cdot x^3$. What existing strip does it match for all values of x?

14. Make a conjecture for $x^a \cdot x^b$. Test your conjecture by constructing another product of two powers.

15. Now go to page "Division." You'll use this page to explore division problems.

16. Use the **a/b** custom tool to construct a strip for $\frac{x^7}{x^3}$. What is the result?

17. Now find $\frac{x^8}{x^2}$. What is the result?

18. Make a conjecture for $\frac{x^a}{x^b}$. Test your conjecture by constructing another quotient of powers.

19. Go to page "Powers." You'll use this page to explore problems like $(x^4)^2$.

 20. Use the **(a)^b** custom tool to construct a strip for $(x^4)^2$. What is the result?

21. Now find $(x^3)^3$. What is the result?

22. Make a conjecture for $(x^a)^b$. Test your conjecture by constructing another power of powers.

EXPLORE MORE

23. Go to page "Explore More." What if the bases are not the same? Is there a rule for problems like $x^a \cdot y^b$? Experiment and then describe your conclusions.

3

Functions

Function Machines: Introducing Functions

Students investigate how to determine the hidden one-step rules used by virtual function machines, and come to understand that the output of a rule depends on the input. Along the way, students explore questions such as these: What are good inputs to try? How many must be tried? Does it help to keep track of the inputs and outputs? The input/output pairs are collected in a table on screen. Optionally, the table's data may be graphed as well.

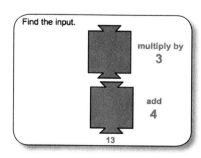

Function Machines: Equivalent Expressions

Students operate function machines, collect data, and look for a machine's rule. They discover that more than one rule can produce the same outputs for a set of inputs, and look for an explanation. Students may also graph data they have collected in a table.

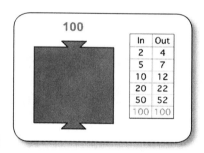

Function Machines: Working Backward

Students use interactive function machines to explore the reciprocal nature of addition and subtraction and of multiplication and division. Given outputs, students determine the unknown inputs by working backward, undoing each operation of a machine's two-step rule.

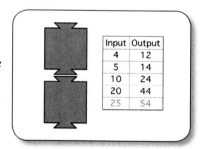

The Function Game: Determining Linear Equations

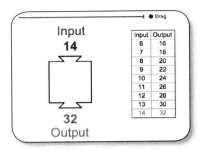

Students guess the rules relating a linear function's input and output. Then they create a rule to challenge their peers. As a result, they begin to understand the meaning of the slope and y-intercept parameters in a linear equation. Students see how the output is calculated by multiplying the input by the constant-change value and then adding the starting value.

Function Machines: Introducing Functions  ACTIVITY NOTES

INTRODUCE

Project the sketch for viewing by the class. Expect to spend about 20 minutes.

1. Open **Function Machines Introducing.gsp.** Go to page "Machine 1." Make sure the input showing is 2. To change an input, double-click it, enter a new value in the dialog box that appears, and click **OK.**

2. Explain, ***This is an in-out machine. A number will go in when I press*** **Run Machine.** ***What number is about to go in?*** [2] ***What do you think will happen when I press*** **Run Machine?** Take responses and then press the button. ***What number came out of the machine? What do you think the machine did to change the "in" value of 2 to the "out" value, 4?***

   Encourage students' creativity. Students don't know yet that this is a one-step machine, so they may suggest possibilities such as *It multiplied the 2 by 4 and then divided by 2.* If students' ideas don't include the two possibilities that the machine added 2 to the input and that it multiplied the input by 2, offer these ideas. Record all suggestions on the board in the form, "multiply by 4, divide by 2" and "multiply by 2."

3. ***Here's something else I can tell you about this machine: It only carries out*** one ***operation on the number that goes in. (We'll call that number the*** input.*) For example, it might multiply the input. It doesn't do one operation, such as add, and then do another, such as multiply.*

   We'll call what the machine does its rule. ***Now that you know that the machine has only one rule, which of the rules we've written down should we keep?*** Have students identify "add 2" and "multiply by 2."

   Are you sure both of these rules could work for the input of 2 and output of 4? Though the answer may seem obvious, this is an important question; be sure to give all students enough time to consider it. ***But the machine has only one rule! How do you think we can find out what that rule is?*** Have students discuss this in pairs before taking responses.

   Some students may suggest trying another input, but they may not be ready to explain why that would be helpful. Other students may want to make the case that the class will need to test one or more new inputs because another input could give an output that would fit one of the rules, but not the other. ***Give us an example of what you mean.*** A sample student response is this: *Let's say we make the input 5. If the rule is "add 2," we know a 7 will come out of the machine. But if the rule is "multiply by 2," we know a 10 will come out of the machine.* ***Who agrees with this idea?***

All students may not be convinced at this point. This is fine. The idea will be developed throughout the activity.

DEVELOP

Expect to spend about 25 minutes on this part of the activity. Continue to project the sketch.

4. Before the class embarks on trying other inputs and generating sets of input/output pairs, ask, ***How might we keep track of the inputs we try and the outputs the machine produces?*** Elicit the idea of making a table. Distribute the worksheet *In-Out Machines* and direct students' attention to Table A. Have students record 2 and 4, respectively, in the first row. In the sketch, create a table by selecting the input and output, in that order, and choosing **Number | Tabulate**. Double-click the table to enter the data.

It is possible to enter values for "*In*" that extend into the tenths place or beyond. However, the "*In*" and "*Out*" values on screen are set to display only to the nearest whole unit.

5. ***Let's try another input.*** Press *Reset* to prepare the machine to run again. Ask a volunteer to suggest a *whole number* for the input. Change the "*In*" parameter to this new value.

   Ask students to predict the output for each of the rules the machine may be using: "add 2" and "multiply by 2."

   Run the machine again. Have students record the pair in the second row of Table A. Double-click the table to add the resulting input/output pair.

6. ***Now that you have another input and output, what do you think?*** Facilitate discussion. Record ideas on the board and encourage students to debate them. Notice the ideas they have about whether they can predict with more certainty which of the two rules the machine is using. How many inputs do they think it's necessary to try in order to be certain of the rule the machine is using?

To save time, you can change the "*In*" value without first pressing *Reset*.

7. Run the machine several more times. Each time, encourage discussion about good numbers to use as inputs. Students may discover that small numbers, multiples of ten, and other "easy" numbers are simpler to work with than larger numbers. Students should record all input/output pairs in Table A as you double-click the Sketchpad table to add them.

8. When Table A is complete, ask students to study the inputs and outputs and write the rule they think the machine is using. Facilitate discussion. ***Are you convinced you know what the rule is? Why or why not?*** When most students are convinced they know the rule, press *Show/Hide Rule*. Does the rule shown, "add 2," confirm or challenge students' thinking?

New Machines, New Rules

9. Go to page "Machine 2" and facilitate as the class works to find the rule for this machine. The rule is now "subtract 3." Have students record inputs and outputs in Table B as you record in a Sketchpad table.

10. Go to page "Machine 3" and facilitate as the class works to find the rule for this machine. The rule is "multiply by 4." Have students record inputs and outputs in Table C as you record in a Sketchpad table.

SUMMARIZE

Expect to spend about 30 minutes on this part of the activity. Continue to project the sketch.

11. Have the class create problems of the type they have been solving—a good technique for helping them consolidate their understanding. *Now you get to give the machines their rules! You can change the number associated with each rule, but not the operation.* Using Machine 2, model these steps for changing a rule.

   • Show the rule.

   • Have a volunteer cover the projector lens.

   • Have another volunteer change the value of the "*In*" parameter and press *Show/Hide Rule.*

Have students make rules for the class to determine. Each rule maker can act as moderator at the computer, taking suggestions for inputs and running the machine. Students can create tables to keep track of input/output pairs.

Encourage creative challenges, such as adding 0 or multiplying by 0 or 1.

12. Engage the class in shared writing to help students synthesize what they have experienced. *Let's make a poster that tells what you've learned about in-out machines. What can you say about how in-out machines work? What can you say about determining the rule a machine is using?* Here are three big ideas that should emerge from students' experience with in-out machines in this activity.

   • If you know the rule a machine is using, for any input, you know what the output must be. The output for a rule depends on the input.

   • Each input has a single corresponding output. If you run a machine again with the same input, the output has to be the same.

   • To determine the rule, you probably need to know more than one input/output pair.

 **ACTIVITY NOTES**

EXTEND

1. The work in this activity may have prompted new queries by students. Record the questions posed and provide time for students to investigate them. Here are some examples.

   How many numbers do we need to try before we know the rule for sure? Are two numbers enough?

   What if the rule were division?

   Could a machine have a rule other than add, subtract, multiply, or divide by a number? [For example, what if the machine squared the input?]

   Could a machine have two rules? How hard would it be to determine a two-step rule?

2. For students who would benefit from more individualized work with the sketch, consider giving them an opportunity to use **Function Machines Introducing.gsp** at a later time. Pairs of students can create and share in-out machine challenges with each other by changing the numbers associated with the rules.

3. Graph the input/output data you have collected in a table.

   • Select the table and choose **Edit | Copy.**

   • Add a blank page using **File | Document Options.**

   • Choose **Edit | Paste.** Hide any parameters that are showing.

   • Select the table and choose **Graph | Plot Table Data.** Click **OK.**

   • To see more points, drag the origin and change the scale of the grid.

Exploring Expressions and Equations in Grades 6–8 with The Geometer's Sketchpad
© 2012 Key Curriculum Press

In-Out Machines

Look for the rule an in-out machine is using.

Table A

In	Out

Rule: _____

Table B

In	Out

Rule: _____

Table C

In	Out

Rule: _____

Table D

In	Out

Rule: _____

Function Machines: Equivalent Expressions  ACTIVITY NOTES

INTRODUCE

Project the sketch for viewing by the class. Expect to spend about 15 minutes.

1. Introduce the Sketchpad model. Open **Function Machines Equivalent.gsp** and go to page "Example." Invite students to predict what will happen when the machine is run.

   Press *Run the Machine*. The input 4 will enter the top and the output 18 will exit the bottom. Ask, ***What two-step rule could the machine be using?*** Take responses.

   Press *Reset*. Change the input by double-clicking it and typing a new value in the dialog box that appears. ***Predict the output.*** Do this several times until students are comfortable with the function machine representation. As the class works, incorporate the terms *input value* and *input*, and *output value* and *output*.

Discovering More Than One Rule

Note that Machine 1 displays outputs to the nearest tenth. This works fine for the "divide by 2" step used by the machine now, but will be problematic if you later decide to use a rule like "divide by 3."

2. Go to page "Machine 1." ***Now the rule used to transform inputs into outputs is hidden.*** Press *Run the Machine*. The input 2 yields an output 11. Create a table by selecting the input and output, in that order, and choosing **Number | Tabulate**. Double-click the table to enter the data.

   Ask the class to propose rules the machine might be using. Record responses. Some possibilities are "multiply by 2, add 7;" "multiply by 4, add 3;" and "add 20, divide by 2." If students propose the rule "add 9," remind them that the rule consists of two parts, one applied by the top of the machine and the other by the bottom of the machine. Some students might then split this rule into "add 3, add 6."

   Point to one of the students' proposed rules. ***How confident are you that this is the rule the machine is using?*** Because more than one of the proposed rules "works," most students will agree that they cannot have confidence in any of their rules when only one input/output pair is available.

If you forget to press *Reset* before entering a new input, the resulting output will appear immediately. You can think of this as a shortcut to use if you wish to speed up the process of obtaining input/output pairs to enter in the table.

3. Invite the class to suggest another whole-number input. Allow time for students to consider and discuss what numbers are good to use. Press *Reset*, enter the new input, and run the machine again. (Note that the displayed outputs are rounded to the nearest tenth.)

   Double-click the table to add the resulting input/output pair.

   Repeat this process with several more inputs.

Exploring Expressions and Equations in Grades 6–8 with The Geometer's Sketchpad

4. Ask the class to study the table of input/output pairs. ***Do any of the rules you proposed describe how the outputs are obtained from the inputs?*** If none of the rules work, ask students to suggest other rules the machine might be using.

   Press *Show the Rule* when most of the class feels confident in their prediction of the rule. The machine uses the rule "divide by 2, add 10." Some students, however, may have predicted "add 20, divide by 2." Note that each of the two rules seems to work: they both produce the same set of input/output pairs in the table. ***Are you convinced that two rules can produce the same outputs?*** Take responses. This question is intended to lead into the next part of the activity.

DEVELOP

Continue to project the sketch. Expect to spend about 20 minutes.

5. ***Let's look at another machine.*** Distribute the worksheet *Do We Know the Rule?* Go to page "Machine 2" of **Function Machines Equivalent. gsp.** The number 5 is set to run through the machine. Press *Run the Machine.* The output is 14. Select the input and output, and choose **Number | Tabulate.** Double-click the table.

6. Direct students' attention to the tables on the worksheet. Explain that Tables A–C show three rules another class proposed after seeing the input/output pair (5, 14). Also, the tables list other inputs the class tried. (They chose multiples of 5 to make dividing by 5 easier in Table C.) Give students time to confirm that all three rules work for the input/output pair (5, 14).

   Have students complete all three tables, using mental computation to find the outputs for the given inputs.

7. Ask the class how they can determine whether any of the rules in Tables A–C work for the machine shown on screen. Take responses and then run each input in the tables (10, 100, 15, and 25) through the machine. Enter each input/output pair in the table on screen.

   When all the inputs have been run, ask, ***What do you know about the rules for Tables A through C now?*** Students should see that the outputs in Tables A and C do not match the machine's outputs. The outputs in Table B do.

   How confident are you that Table B's rule is the one the machine is using? Invite discussion, and then press *Show the Rule.* The machine

shows a two-step rule different from the two-step rule for Table B: "multiply by 2, add 4."

8. Have students write the machine's two-step rule under Table D and confirm, using mental computation, that the machine's rule works for the five inputs.

SUMMARIZE

Continue to project the sketch. Expect to spend about 25 minutes.

9. Have students work on worksheet steps 2–4 and then bring the class together to discuss them.

10. Now have students work on steps 5–7, and again bring students together for discussion. ***Can you figure out why all three of the rules produce the same input/output pairs?*** Allow plenty of time for students to explain to each other their ways of reasoning about the rules. Students may not explicitly identify the distributive property at work, but they may offer explanations that show an intuitive understanding and application of the property.

11. Invite students to study two rules that both give the same outputs for a set of inputs: "multiply by 2, add 4" and "add 2, multiply by 2." Can students explain why these rules produce the same input/output pairs? Alternatively, provide this question as a writing prompt and have students respond individually. Here are sample student responses.

If you add 2 and then double it, that's the same as adding 4. It doesn't matter what the input is. If you're always adding 2 and doubling that, you're always adding 4 to twice the input.

$$\text{input} + 2$$

$$2 \times \text{input} + 2 \times 2$$

$$2 \times \text{input} + 4$$

If you add 2 to the input, then, in the next step, when you multiply by 2, that 2 will become 4. Let's say the input is 3. First, you add 2. Then, when you multiply by 2, you're multiplying the 3 by 2, and you're multiplying the 2 by 2. So, it's the same as multiplying the input (3) by 2 and then adding 4.

$$3 + 2$$

$$2 \times 3 + 2 \times 2$$

$$2 \times 3 + 4$$

12. ***What can you say about predicting the rule for a machine? Can you know for sure the rule the machine is applying?*** The discussion should elicit the ideas expressed in the sample student responses here.

It's possible for more than one rule to produce the same set of input/output pairs.

Because there can be more than one rule, we can predict a rule the machine might be using, but we can't know for sure. The only way to know for sure is to have the machine report the rule.

EXTEND

Remind students that they can change only the numbers, not the operations.

1. For students who would benefit from additional practice, have student pairs use page "Machine 2" to create function-machine challenges. The challenger changes the rule by pressing *Show the Rule*, double-clicking both numbers associated with the rule, and changing their values, and then pressing *Hide the Rule*. The solver records input/output pairs in a table and predicts a rule the machine could be using.

2. Page "Extend" presents a pair of function machines. The machines run simultaneously, using the same number as an input. Students should choose several different inputs, run the machines, and note that the outputs are always the same for both machines. Ask students to identify two different rules the machines could be using to yield the same input/output pairs.

Challenge students to change the rules associated with the machines to create a new pair of machines that have different rules but produce the same output for any input.

3. Graph the input/output data you have collected in a table.

   • Select the table and choose **Edit | Copy.**

   • Add a blank page using **File | Document Options.**

   • Choose **Edit | Paste.** Hide any parameters that are showing.

   • Select the table and choose **Graph | Plot Table Data.** Click **OK.**

   • To see more points, drag the origin and change the scale of the grid.

 ACTIVITY NOTES

ANSWERS

1.

Table A	
in	**out**
5	14
10	29
100	299
15	44
25	74

multiply by 3
subtract 1

Table B	
in	**out**
5	14
10	24
100	204
15	34
25	54

add 2
multiply by 2

Table C	
in	**out**
5	14
10	15
100	33
15	16
25	18

divide by 5
add 13

Table D	
in	**out**
5	14
10	24
100	204
15	34
25	54

multiply by 2
add 4

2. All four rules apply.

3. Only the third rule applies: multiply by 3, subtract 5.

4. Answers will vary. Students may suggest trying other input values in order to check that the third rule applies.

5. All four rules apply.

6. Only the third rule applies: multiply by 2, divide by 5.

7. Answers will vary.

Do We Know the Rule?

 Name:

Explore rules for function machines.

1. Use the tables to work with Machine 2.

Table A	
in	**out**
5	14
10	
100	
15	
25	

multiply by 3

subtract 1

Table B	
in	**out**
5	14
10	
100	
15	
25	

add 2

multiply by 2

Table C	
in	**out**
5	14
10	
100	
15	
25	

divide by 5

add 13

Table D	
in	**out**
5	14
10	
100	
15	
25	

_____

_____

2. Vicki and Ramona are using a two-step function machine.
   Their first input/output pair is 6, 13.
   Circle any of these rules that the machine might be using.

multiply by 2	subtract 3	multiply by 3	multiply by 4
add 1	add 10	subtract 5	subtract 11

3. The girls' second input/output pair is 100, 295.
   Which of the rules might the machine be using? Circle one or more.

multiply by 2	subtract 3	multiply by 3	multiply by 4
add 1	add 10	subtract 5	subtract 11

4. What would you do next if you were Vicki and Ramona?

_____

5. Now Vicki and Ramona are using another function machine.
   Their first input/output pair is 10, 4.
   Circle any of these rules that the machine might be using.

divide by 2	multiply by 2	multiply by 2	divide by 5
subtract 1	subtract 16	divide by 5	add 2

Do We Know the Rule?

continued

6. Their second input/output pair is 5, 2.
   Circle any of these rules that the machine might be using.

divide by 2	multiply by 2	multiply by 2	divide by 5
subtract 1	subtract 16	divide by 5	add 2

7. The girls predict that the machine may be multiplying by 4 and dividing by 10. They think of two other rules, also shown here. Can you explain why all three rules work? Answer here and on the back of this sheet.

multiply by 4	multiply by 8	multiply by 10
divide by 10	divide by 20	divide by 25

Exploring Expressions and Equations in Grades 6–8 with The Geometer's Sketchpad
© 2012 Key Curriculum Press

Function Machines: Working Backward  ACTIVITY NOTES

INTRODUCE

Project the sketch for viewing by the class. Expect to spend about 15 minutes.

1. Distribute paper to each student and introduce the Sketchpad model. Open **Function Machines Working Backward.gsp** and go to page "Mystery Input 1." Explain, *A number we'll call the input is hidden behind a ball. When the input enters the machine, it will be multiplied by 3, and then 4 will be added to the result.* Press *Run the Machine.* The red ball will drop and an output of 13 will emerge. (The jiggling is a visual cue that the machine is working.)

2. *We don't know what input went into the machine. But we do know the output. Can you figure out the input?* Have students work on their own or in pairs for a few minutes to figure out the unknown input, and then have them share their methods. Here is a sample student response.

Students' methods will become more refined in the Develop section that follows.

I wasn't sure what the input could be, so I just picked a number and tried it. I started with 4, but that made an output of 16. That was too big, so I tried an input of 2. That was too small. I tried an input of 3—in between—and that worked.

Press *Show Input* to reveal the input.

DEVELOP

Continue to project the sketch. Expect to spend about 20 minutes.

3. Present problems with new mystery inputs in order to give students experience finding the inputs. For each problem, press *New Input.* (The computer will assign a random value from 0 to 12 for the input.) Run the machine again. Ask students to figure out the value of the input. (It's possible that the sketch will pick the same input twice. If it does, simply press *New Input* again.)

As students solve the problems, note whether anyone is working backward.

I thought I could go backward. The output this time was 10. That means some number became 10 when 4 was added to it. That must be 6. Six is some number multiplied by 3. That number has to be 2. So, the input should be 2. I checked by seeing whether 2 times 3, plus 4, equals 10. It does!

If this method is not proposed, ask guiding questions. *How can we figure out what number entered the "add" machine?* Elicit the idea that, because the machine added 4 to some number to obtain the final

output, we need to subtract 4 from the output to find out what went into the "add" machine.

Continue this reasoning by asking about the first machine. *How can we figure out the "multiply" machine's input now that we know its output?* Because the input was multiplied by 3 to get the output, the output must be divided by 3 to find the input.

Let's check. Imagine the number you think is the input goes into the machine. Apply the two-step rule. Do you get the output?

4. Ask students to describe the two-step rule for finding the input when given the output. For the example above, the rule is "subtract 4, divide by 3."

Solving Problems with Different Rules

5. Now create problems by changing the numbers associated with the two-step rule. Double-click a number with the **Arrow** tool, enter a new number in the dialog box, and click OK. In this manner you can change the "multiply by 3, add 4" rule to another rule, such as "multiply by 2, add 5." (Note that you can change only the numbers, not the operations.)

6. Have students solve problems with different operations by using the machines on pages "Mystery Input 2" and "Mystery Input 3." Have students describe a two-step rule for finding the input. (Note that the "Mystery Input 3" machine displays outputs to the nearest tenth. This works fine for "divide by 2," but will be problematic if you choose a rule like "divide by 3.")

SUMMARIZE

Expect to spend about 20 minutes.

7. Ask what students notice about how they have determined the inputs. Here are sample student responses.

We used the opposite operations to undo what the machine was doing. When it added, we subtracted. When it multiplied, we divided.

Our rules are upside-down from the machine's rules. Say the machine starts with multiplying and ends with adding. We start with subtracting and end with dividing.

Elicit the ideas that (1) addition and subtraction "undo" each other, and (2) the strategy students have been using is called working backward, or "undoing."

8. Have students create their own missing input problem to share with friends or family. Distribute the worksheet *Mystery Input* and have students enter their own two-step rule in the blank spaces next to the machine. They should also enter several outputs into the accompanying input/output table, leaving the input column blank.

Discuss with students how they will describe the missing input problem when presenting it to a friend or family member. Explain that students should not offer strategies for solving the problem at first.

Have the class report back on the solution strategies they observed others use to solve the problem. Highlight students' accounts of the working backward (or "undoing") strategy.

EXTEND

1. For students who would benefit from more individualized work on undoing function rules, provide opportunities to work with any of the three Mystery Input machines.

2. Pairs of students can also use the three machines to create challenges for each other. As one student looks away, the other student presses *Show Input*. She double-clicks the value of the input, enters a new value in the dialog box, and clicks OK. Finally, she presses *Hide Input* to conceal the input, and challenges her partner to determine it.

Mystery Input

  Name:

Make a machine for someone else.

1. Write a rule such as "add 4" or "multiply by 3" in each of the blanks.

2. Choose some inputs and find their outputs. Write *only* the outputs in the table. (Try to use mental math to do this.)

3. Share this problem with a friend or family member. Can he or she figure out the inputs?

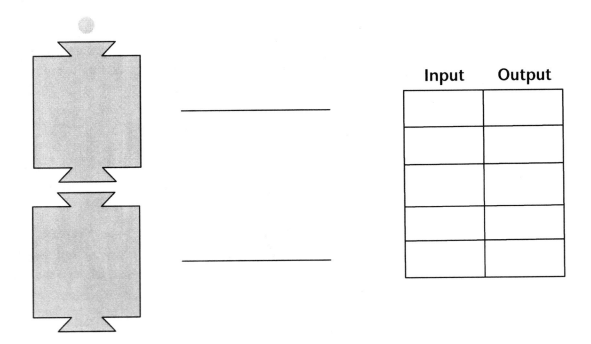

Input	Output

Exploring Expressions and Equations in Grades 6–8 with The Geometer's Sketchpad
© 2012 Key Curriculum Press

The Function Game: Determining Linear Equations

ACTIVITY NOTES

INTRODUCE

Expect this part of the activity to take about 10 minutes.

1. **This activity is about function machines. What are some examples of machines?** Students may suggest quite a variety, such as vending machines, televisions, or cars. **What do these machines have in common?** Elicit the idea that each takes some kind of material or data or action, called *input,* and produces something, called *output.* **What are the input and output for the machines you listed?** For example, a vending machine takes money and your choice as input and produces the item you bought as output. A television's input is the signal for the channel, and its output is its display. A car's input can be thought of as gas, or the sequence of actions needed to drive it, and its output is motion. **Each machine also has a rule for producing the output. What might some of those rules be?** The vending machine's rule, for example, is something of the form, "If you've put in enough money and pressed this button, then I'll slide out one item from that compartment." For a television, the channel you select will determine the display. A car's rule is more complex because of the many inputs.

2. **The machine in this activity is called a function machine. When we say, "My grade is a function of how much homework I do," what do we mean by the word function?** Elicit the idea of one number depending on another. Introduce the idea of a function machine, in which the output number depends on the input number according to some rule. **In this activity you will find the rules being used by some function machines, seeing only their input and output numbers.**

DEVELOP

Expect this part of the activity to take about 20 minutes.

3. Assign pairs to computers and tell them where to locate **Function Game.gsp.** Distribute the worksheet. Ask students to work through step 10.

4. As you circulate, you may notice these things.

   - Some students may be unable to slide the slider. Be sure they have the **Arrow** tool selected. Otherwise, let pairs work at their own pace without interruption.

   - In worksheet step 6, the description of the rule can be in words. Do not insist at this point on any particular form or even on complete accuracy. The purpose of using words here is to motivate students to use compact symbols in worksheet step 7.

 ACTIVITY NOTES

- In step 3, students who have described a rule in some other form might be puzzled about how to write an equivalent rule in this standard linear form. Encourage independent thinking, but if the frustration level becomes too high for learning, ask questions. ***Can you see from the table how to rewrite the rule in this standard form? How much does the output change each time the input increases by 1? How much would the output change if the input increased from 0 to 5? In general, how can you describe the change in the output as the input increases from 0?***

- In worksheet step 9, the machine rules are created using random integers. Because the numbers may be negative, be prepared to review rules of integer arithmetic. ***What do you remember about adding negatives to positives? About multiplying negatives by negatives?***

- Some students may generate rules that describe the relationships between the input and the output for some but not all pairs in their table. Postpone any corrections until Check In.

- Students' methods in worksheet step 10 may vary, and their descriptions may be somewhat vague. You need not clarify them now, but note to yourself how they might be improved during the full-class discussion.

5. Check in with the class after students have finished worksheet step 10. Have a volunteer display a table for one function, but without the rule. ***What is a rule for this function? Do others agree? Why not?*** Out of the class suggestions, bring out these points.

   - A rule for a function machine will describe all, not just some, input/output pairs.

   - A rule can have a variety of algebraic expressions. Expressions for the same rule are called *equivalent expressions.*

   - The standard expression this machine uses has form

     $Output = (\text{common difference}) \cdot Input + (\text{output when input is } 0)$

6. Ask students to continue with the rest of the worksheet. Tell them which students they'll be challenging. You might have already-existing pairs of students challenge each other, each pair challenge another pair, or each student challenge the next one in alphabetical order.

7. As you circulate, continue to let students work independently. Worksheet step 9 is intended to be answered with a yes or a no. Explanations for this answer can wait until the summary.

SUMMARIZE

Expect this part of the activity to take about 15 minutes.

8. Ask students to share any methods they used other than those that arose during Check In, and help the class critique them. Keep asking questions to help clarify the wording. ***What do you mean when you say that number? Does it matter which term you subtract from which other term? Is a good term for that result the* common difference?** Record one or more edited methods on the display or board. One method, for example, might be, *Multiply input by the common difference and then add the output that corresponds to input 0.* Another might be, *Add the start value to the product of the input and the constant-change value.*

9. ***Why do these methods work?*** From the class's suggestions, elicit the idea that repeated addition of the constant difference (change amount) is the same as multiplication by it, because multiplication is repeated addition.

10. ***Do you know what functions with rules like these are called?*** [They are *linear functions,* because their graphs are straight lines.]

11. ***Some people prefer to think of the rule as***

    Output = (output when input is 0) + (common difference) · *Input*

    How could that form be interpreted? [The output that corresponds to input 0 is the starting value. Then you add or subtract the common difference as many times as the input value.]

ANSWERS

5. Tables will vary.

6. Multiply the input by 2 and then add 4. An equivalent rule is to add 2 and then multiply by 2.

7. $Output = 2 \cdot Input + 4$

9. Tables and rules will vary randomly, but they should be consistent with each other. That is, *Output* should equal (common difference) times *Input* plus (output when input is 0).

10. Methods and their descriptions may vary. One is to find the common difference between consecutive input values and find the output value when the input value is 0. The rule will then be

$$Output = (\text{common difference}) \cdot Input + (\text{output when input is 0})$$

11–14. Rules that students create to challenge each other will vary considerably. So might the methods used to guess those rules.

The Function Game

 Name:

In this activity you'll guess rules of functions from pairs of input and output values.

EXPLORE

1. Open the sketch **Function Game.gsp** and go to page "Guess Rule." The *Input* goes in the top, and the function machine changes it into the *Output*, which comes out the bottom.

2. Across the top of the screen is a slider. Drag the blue point and the values of *Input* and *Output* change. Your goal is to find the rule the function machine is using to change the input values into the output values.

3. You'll add pairs of values to the table in the sketch to help you figure out the function rule. Select the *Input* and *Output* values in order and double-click the table to enter them.

4. Drag the blue point to change the values, select them in order, and double-click the table again. Keep adding pairs of values to the table until you can tell what rule the machine is following.

5. Copy your table.

Input	Output

6. Describe the rule the machine is following.

7. You can express the rule in the form of an equation by replacing the boxes with numbers.

Output = ☐ • *Input* + ☐

What is your rule in this form?

8. Select the table and choose **Number | Remove Table Data**. Select **Remove all entries** and click **OK**.

9. Press *New Challenge* to change the rule and repeat this process. When you have found the rule, copy the table and write the rule in the form of an equation.

Input	Output

10. Keep changing the rule and making tables until you see a method to determine the function's rule. Describe your method.

11. Now you will get to challenge someone else by making your own rule. Go to page "Create Rule" This time the rule is also displayed. Double-click each red number and enter new numbers to change the rule. Then press *Hide Function Rule.*

    What rule did you set up?

12. Challenge someone else to guess the rule. Who took your challenge?

13. What method did this person use in trying to guess the rule?

14. Was this method successful?

Exploring Expressions and Equations in Grades 6–8 with The Geometer's Sketchpad
© 2012 Key Curriculum Press

Graphs of Linear Functions

Fly on the Ceiling: Coordinate Systems

Students name ordered pairs on a coordinate plane, add the coordinates to tables, and plot points from coordinates in a table. They solve a series of coordinate geometry puzzles in which they find coordinates for missing vertices of simple geometric shapes.

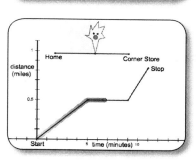

Filling Gaps: Tables, Graphs, and Equations

Students are challenged to fill in gaps in a linear table. They plot points corresponding to the known values and guess at function expressions until they plot a line that goes through those points. Then they watch the coordinates of a moving point on that line to determine the missing values.

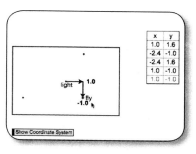

Mellow Yellow: Interpreting Graphs

Students interpret linear piecewise time-distance graphs that represent different stories about a character, Mellow Yellow. They decide whether a given graph corresponds to the motions (walking fast, walking slow, stopping, going backward) described in the story. Students then create stories based on given graphs and create graphs based on given stories.

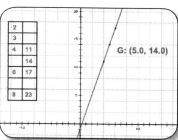

Weather Balloon: Comparing Distance and Speed

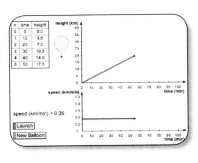

Students study how a table and graph can represent heights of a rising weather balloon and distinguish between the representations of distance and speed. They calculate and plot points for the speed, and they contrast the graphs of height and speed with respect to time. Students see that speed is reflected in the amount a line rises per unit of time.

Dance Pledges: Plotting Linear Equations

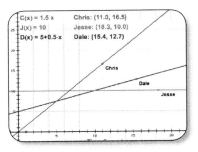

Students plot linear equations that represent different plans for a dance marathon fundraiser. They investigate relationships between the equations (in function form) and the slopes and y-intercepts of their graphs, and then use sliders to dynamically change the slopes and y-intercepts.

Hikers: Solving Through Multiple Representations

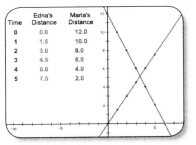

Students use tables, graphs, and equations to represent and solve a real-world problem about two hikers walking at different speeds in opposite directions along the same trail.

Car Wash: Using a Graph to Make Decisions

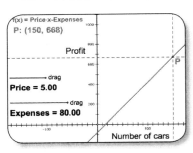

Students determine, graph, and interpret an equation of a line modeling profit from a car wash fundraiser. They vary conditions in the context and observe the corresponding changes in the graph.

Exploring Expressions and Equations in Grades 6–8 with The Geometer's Sketchpad
© 2012 Key Curriculum Press

Fly on the Ceiling: Coordinate Systems

 ACTIVITY NOTES

INTRODUCE

Project the sketch for viewing by the class. Expect to spend about 10 minutes.

1. Open **Fly on the Ceiling.gsp.** Go to page "Ceiling." Enlarge the document window so it fills most of the screen.

2. Explain, *Today you are going to use Sketchpad to explore a coordinate plane as mathematician and philosopher René Descartes first saw it—as a fly crawling on a ceiling.* Recount the story about Descartes lying in his bed and realizing that he could describe a fly's position on the ceiling by using two numbers that described its distance from each of two walls.

Descartes is perhaps better known as a philosopher than as a mathematician. His most famous quote is *Cogito ergo sum*—"I think, therefore I am."

3. Model worksheet steps 1 and 2 for the class. Here are some tips.

- In worksheet step 1, model dragging the point. ***What happens to the measurements on the sketch and in the table?*** [The measurements change as the point moves.] *The measurements are called* ***coordinates.*** *A coordinate is one of the two numbers that locates a point on a coordinate plane. The coordinate plane in this case is the ceiling.*

- Review with students that it takes two coordinates to locate a point on a coordinate plane. ***Suppose I tell you the fly is 2 meters from the left side of the ceiling. Can you tell me exactly where the fly is?*** [No, students will only be able to state that the fly is somewhere 2 meters from the left side.] ***What other information do you need to find the fly's location?*** [How far it is from the bottom (or top) side]

- Make sure students understand that the numbers in the table correspond to the metric measurements from the two walls to the point. Drag the point to (2, 1). ***How far is the fly from the left side of the ceiling?*** [2 meters] ***How far is the fly from the bottom side of the ceiling?*** [1 meter] ***What do the numbers in the table tell you?*** [The *x*-value is the distance from the left side; the *y*-value is the distance from the bottom side.]

- Now drag the point to (1.3, 2.4). ***How far is the fly from the left side of the ceiling now?*** [1.3 meters] ***From the bottom?*** [2.4 meters] Be sure students understand that the measurements are in tenths of a meter. ***How are the distances measured?*** [In tenths of a meter]

- In worksheet step 2, the terms *x*- and *y*-coordinates are introduced as names for the two coordinates. Tell students that (0, 0) is known as the *origin* of a coordinate plane. ***The coordinates tell the distance***

from the origin. The x-coordinate tells the distance left or right from the origin. The y-coordinate tells the distance up or down from the origin. In which corner is the origin? [Bottom-left corner] Drag the point to the bottom-left corner to show students the origin.

- Now drag the point to the top-right corner. *How far is the fly from the left side of the ceiling?* [6 meters] *How far is the fly from the bottom side of the ceiling?* [4 meters] *How could we name the location of this point?* Students may suggest (4, 6) or (6, 4). *There is a standard way of referring to points, so there isn't any confusion about the location. The left and right, or horizontal, distance from the origin is listed first. This is the x-coordinate. The up and down, or vertical, distance from the origin is listed second. This is the y-coordinate.* Explain that the two coordinates for the point in the top-right corner are (6, 4).

- Demonstrate how to double-click the table to enter a measurement.

....

Students can double-click the table with the **Arrow** tool while holding down the Shift key to remove the most recently added value.

4. *Now you will explore Descartes' coordinate plane on your own. As you work, think about how you can use coordinates to find locations and to determine measurements.*

5. If you want students to save their work, demonstrate choosing **File | Save As,** and let them know how to name and where to save their files.

DEVELOP

....

Expect students at computers to spend about 35 minutes.

6. Assign students to computers and tell them where to locate **Fly on the Ceiling.gsp.** Distribute the worksheet. Tell students to work through step 22 and do the Explore More if they have time.

7. Let pairs work at their own pace. Encourage students to ask their neighbors for help if they are having difficulty using Sketchpad. As you circulate, here are some things to notice.

- In worksheet step 3, some students may not drag the point completely into the corner. Tell students to drag the point as far as possible in both directions. The coordinates will be whole numbers.

- In worksheet step 4, listen to students as they discuss the answer to this problem. You may hear a range of comments that can inform your teaching. Here are some sample student responses.

  The two right corners are both 6 meters from the left side of the ceiling. The two top corners are both 4 meters from the bottom side of the ceiling. That makes the dimensions 6 meters by 4 meters.

Exploring Expressions and Equations in Grades 6–8 with The Geometer's Sketchpad
© 2012 Key Curriculum Press

Look at the coordinates of the top-right corner, (6, 4). The 6 tells us that the ceiling is 6 meters long. The 4 tells us the ceiling is 4 meters wide.

If we drag the point along the bottom edge toward the far right, we can read the number after the arrow. That is the length. Then let's drag the point up along the right side to the top-right corner. That number is the width.

• In worksheet steps 5 and 6, if students have trouble finding the center, have them think about one dimension at a time. **How would you find the center of the horizontal distance? How would you find the center of the vertical distance? What are the coordinates of a point that is at the center of both?**

Have students drag the fly into a corner so they can see the points more easily.

• In worksheet steps 8–10, students should recognize the vertices of each shape. Encourage students who say, for example, just "a square" to write more in their descriptions. **What else can you say about the shape? For example, can you find the length of each side?**

Students can select and drag the measurements to move them next to each point.

• In worksheet step 11, for students who have trouble identifying the points, ask them to recall what the properties of a square are. **What are the properties of a square? How can you use these properties to help you find the coordinates of the missing vertices?**

• In worksheet steps 12 and 13, remind students of the properties of a right triangle and an isosceles triangle, if needed. **How would you describe a right triangle?** [A triangle with a 90° angle] **How can you plot a point to create a 90° angle?** [The point must create two perpendicular sides when the points are connected to form the triangle.] Continue questioning in a similar manner for an isosceles triangle.

• In worksheet step 14, students should have fun swatting Descartes' flies. Watch for students who reverse the *x*- and *y*-coordinates. **Do (0.7, 3.0) and (3.0, 0.7) name the same point? Explain.**

If you do this activity over two days, start the second day here.

• In worksheet steps 15 and 16, students explore the four quadrants on the coordinate plane. By dragging the fly around, they should notice that the *x*-value is negative when the fly is to the left of the origin and the *y*-value is negative when the fly is below the origin. Be sure students understand that the light is the origin. **What are the coordinates of the light?** $[(0, 0)]$

- In worksheet steps 17–19, students will discover that the position of the origin does not make a difference when using coordinates to measure distances. This may be confusing to some students. Stress that all measurements are positive values and that the coordinates tell how far a point is from the origin. ***Can the room's length or width have a negative value?*** [No] ***What does a negative x-value tell you?*** [How far a point is to the left of the origin] ***What does a negative y-value tell you?*** [How far a point is below the origin]

- In worksheet step 21, students find the coordinates for two missing vertices of a rectangle. Ask students to think about the relationship between the missing vertices and the two known vertices. ***How can you figure out where the missing vertices go? What are their positions in relation to the known vertices?*** If students need help, have them press *Show Coordinate System* to see the coordinate plane.

- If students have time for the Explore More, they will have fun plotting points to create designs and puzzles for their classmates to solve.

8. If students will save their work, remind them where to save it now.

SUMMARIZE

Project the sketch. Expect to spend about 15 minutes.

9. Gather the class. Students should have their worksheets with them. Open **Fly on the Ceiling.gsp** and go to page "Ceiling." Discuss worksheet steps 2 and 3. ***We are looking at Descartes' ceiling. The fly moves to the bottom-left corner. How can you use coordinates to tell where the fly is?*** [(0, 0)] Drag the point to that location to verify the coordinates. Double-click the table to enter the measurements. Continue in a similar way for each of the remaining three corners.

10. ***How did you determine the length and the width of the room?*** Have volunteers explain their methods. Make sure students understand that they need to find the range of the *x*- and *y*-coordinates.

11. Discuss how students found the answers to worksheet steps 6 and 7. Students' responses may vary. Here are some sample solutions to worksheet step 6.

    We know the length of the ceiling is 6 meters, so the center of the length would be half of that, or 3 meters. The width of the ceiling is 4 meters. Half of that is 2 meters. We figured the center would be 3 meters over and 2 meters up, at (3, 2).

The x-coordinates for the two bottom corners are 0 and 6. To find the center, we added them and divided by 2 to get 3. The y-coordinates for the bottom and top corners are 0 and 4. To find the center, we added them and divided by 2 to get 2. Our center point is $(3, 2)$.

12. Have students discuss their answers to worksheet steps 9 and 10. Depending on the level of your students, you may wish to discuss finding the perimeter and the areas of the shapes. **What is the formula for the area of a square?** $[s^2]$ **What is the length of one side? How do you know?**

13. Review worksheet steps 11–13, having students share their strategies for finding the coordinates of the missing points.

14. Discuss the four quadrants of a coordinate plane and how the values of the coordinates change in each one. Check for understanding. **In which coordinate would you find the following points?**

    • point A $(-4, 3)$ [Quadrant II]

    • point B $(-4, -3)$ [Quadrant III]

    • point C $(4, 3)$ [Quadrant I]

    • point D $(4, -3)$ [Quadrant IV]

15. In worksheet steps 17 and 18, discuss how students found the dimensions of the room and how they are the same as in worksheet step 4. **Suppose I move the origin to the top-right corner. Do the dimensions of the room change? Explain.** [No, the dimensions are independent of the origin.]

16. In worksheet step 21, have students share their strategies for finding the missing vertices of the rectangle.

17. If time permits, discuss the Explore More. Students can share their designs with the class.

18. You may wish to have students respond individually in writing to this prompt. **A square has side lengths of 3 meters. Explain where the square is for each description of its vertices.**

    • **All vertices have positive coordinates.** [Quadrant I]

    • **All vertices have negative coordinates.** [Quadrant III]

    • **Two vertices have negative x- and y-values and two vertices have positive x-values and negative y-values.** [Quadrants III and IV]

EXTEND

1. Ask students to find other types of triangles on page "Triangles" (worksheet steps 12 and 13) such as isosceles right triangles or equilateral triangles. Placing the third vertex at (3, 3) or (5, 3) forms an isosceles right triangle. Placing the third vertex at (4.0, 2.7) approximates an equilateral triangle.

2. Explore missing vertices of a rectangle further. ***If the opposite vertices of a rectangle have coordinates (a, b) and (c, d), name one possible set of coordinates for the other two vertices using only a, b, c, and d.*** [(a, d) and (c, b)]

ANSWERS

4. The room is 6 meters long and 4 meters wide. These are the ranges of the *x*- and *y*-coordinates.

6. The coordinates of the center of the rectangle are (3, 2). These are half the values of the ranges of the *x*- and *y*-coordinates.

7. The coordinates of the point are (4, 1).

9. The figure is a square with side length 2 m. Its perimeter is 8 m and its area is 4 m^2.

10. The figure is an isosceles trapezoid. The bases are 2 m and 4 m. Depending on the background of your students, they might also calculate that the two other sides are each $\sqrt{5}$ m, the perimeter is $(6 + 2\sqrt{5})$ m, and the area is 6 m^2.

11. The missing coordinates are (3.0, 1.0) and (5.5, 3.5).

12. Answers will vary. There are many points that can be used to make a right triangle. Most of them have an *x*-coordinate of either 3 or 5. Possible solutions using only integers: (3, 2), (3, 3), (3, 4), (5, 2), (5, 3), (5, 4)

13. Answers will vary. There are many points that can be used to make an isosceles triangle. All of them either have an *x*-coordinate of 4 (the average of the two given *x*-coordinates) or have the same *x*-coordinate as one of the given points and a *y*-coordinate of 3. Possible solutions using only integers: (4, 2), (4, 3), (4, 4), (3, 3), (5, 3)

Exploring Expressions and Equations in Grades 6–8 with The Geometer's Sketchpad
© 2012 Key Curriculum Press

 ACTIVITY NOTES

14. The constellation is the Big Dipper, one of the most distinctive constellations in the northern sky.

16. The *x*-coordinate has a negative value when it is left of the origin. The *y*-coordinate has a negative value when it is below the origin.

18. The room is 6 meters long and 4 meters wide. These are the ranges of the *x*- and *y*-coordinates.

19. The dimensions for the room are the same. The size of the room is independent of the position of the origin.

20. In Quadrant I, both coordinates are positive.

    In Quadrant II, *x* is negative and *y* is positive.

    In Quadrant III, both coordinates are negative.

    In Quadrant IV, *x* is positive and *y* is negative.

21. $(-2.4, 1.6)$ and $(1.0, -1.0)$

22. The figure is a regular hexagon.

23. Designs will vary. Check students' work.

24. Designs will vary. Check students' work.

Fly on the Ceiling

One day, philosopher and mathematician René Descartes noticed a fly walking on the ceiling. Descartes realized that he could describe the fly's position on the ceiling by two numbers: its distance from each of two walls. Thus was born the *coordinate plane,* also called the *Cartesian coordinate system* after Descartes. You'll explore the coordinate plane in this activity.

EXPLORE

Open **Fly on the Ceiling.gsp** and go to page "Ceiling." You will see a model of the ceiling in Descartes' bedroom. The red point is the fly in the story.

 1. Drag the fly and notice how the measurements change on the sketch and in the table to the right. These measurements are called *coordinates.* You can also move the fly in very small steps using the Arrow keys on the keyboard.

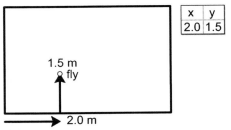

2. Drag the fly to the top-right corner. Double-click the table to enter the measurements of the *x*- and *y*-coordinates.

3. Drag the fly to the other three corners and enter their coordinates into the table. Compare your four pairs of measurements with the measurements of other students.

4. How long and wide is the room? Explain how you can figure this out from the coordinates of the four corners.

5. Drag the fly as close as possible to the center of the room.

6. What are the coordinates for the center point? Describe how you can figure this out using the coordinates of the four corners.

7. If the fly is resting $\frac{1}{4}$ from the bottom edge and $\frac{2}{3}$ from the left edge, what are its coordinates?

 Exploring Expressions and Equations in Grades 6–8 with The Geometer's Sketchpad
© 2012 Key Curriculum Press

CONSTRUCT

Now you will plot points using their coordinates.

8. Plot these four points: (2, 1), (2, 3), (4, 1), and (4, 3). To plot points, choose **Graph | Plot Points.** For each point, enter its coordinates and click **Plot.** After you enter the last point, click **Done.**

9. Describe in detail the figure outlined by the points.

10. Go to page "Mystery Shape." The data in the table represent the last four locations that the fly rested on the ceiling, which happen to form a special shape. Drag the fly to each point given in the table. Can you visualize the shape? Now select the table and choose **Graph | Plot Table Data.** Then click Plot. Describe the figure outlined by the points.

11. Go to page "Missing Corners." The two points are opposite vertices of a square. What are the coordinates for the two missing vertices? Plot them. You can find the coordinates of a point by selecting it and choosing **Measure | Coordinates.**

12. Go to page "Triangles." The two points are two vertices of a triangle. Write down a pair of possible coordinates for the missing vertex so that the triangle is a right triangle. Plot the missing point. Then write down another pair of possible coordinates that would form a right triangle.

13. Delete the point you plotted in the last step. Write down a pair of possible coordinates for the missing vertex so that the triangle is an isosceles triangle. Plot the missing point. How is the *x*-coordinate of the missing vertex related to the *x*-coordinates of the original two points?

Go to page "Fly Swatter." René looked back up at his ceiling and noticed a bunch of flies, so he grabbed his fly swatter to get rid of them. The table lists the coordinates of each fly he swatted. When he looked back up at the ceiling, he noticed that the squashed flies resembled a well-known constellation.

14. Drag the fly to each point listed in the table and press *Fly Swatter.* Then plot the table data to check your work. What is the name of the constellation?

EXPLORE

So far you've seen how Descartes measured the coordinates of a fly by noting its distance from each of two walls. It's possible to use instead a single fixed point, such as a light hanging in the middle of the room, as a reference point.

This reference point is called the *origin* of a coordinate system. So far the origin has been the bottom-left corner, but now you'll use a different origin.

15. Go to page "Ceiling with Light." Notice how the position of the fly is now described by using horizontal and vertical measurements from the location of the light. The light is the origin of a new coordinate system. Drag the fly and notice how the measurements change on the sketch and in the table to the right.

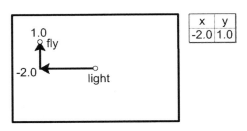

16. What happens to the *x*-coordinate when the fly is to the left of the light? And what happens to the *y*-coordinate when the fly is below the light?

17. Drag the fly to the top-right corner. Double-click the table to enter the measurements of the *x*- and *y*-coordinates. Then drag the fly to the other three corners and enter their coordinates into the table.

18. How long and wide is the room? Explain how you can figure this out from the coordinates of the four corners.

19. Did you get the same dimensions for the room in step 18 as you did in step 4 when the origin was at the bottom-left corner? Does the location of the origin make any difference when using coordinates to measure distances?

20. Press *Show Coordinate System.* The *x*- and *y*-axes divide the plane into four regions called quadrants, numbered I, II, III, and IV, as shown here. For each of the quadrants, state a general rule about the signs of the coordinates of a point in that quadrant.

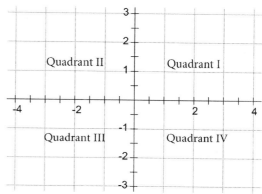

21. Go to page "Incomplete Rectangle." The two points are opposite vertices of a rectangle. What are the coordinates for the two missing vertices? Plot them.

22. Now go to page "Hidden Shape." Look at the table and try to visualize where the points will appear when you plot them. Drag the fly to help you. Can you visualize the hidden shape? Then plot the table data. Describe the figure outlined by the points.

EXPLORE MORE

23. Go to page "Target." Double-click on the values for *x* and *y* and change them to move the target. Then press *Go to Target* to move the arrow that leaves a trace. Once you get the hang of it, choose **Display** | **Erase Traces** to start over and create a design of your own.

24. Go to page "Make Your Own" and create your own design out of points. Drag the fly to each point, double-click the table to add the coordinates to the table, and then delete the point. Have a classmate try to figure out what your design is by looking at the table and then check by plotting the table data.

Filling Gaps: Tables, Graphs, and Equations  ACTIVITY NOTES

INTRODUCE

Project the sketch for viewing by the class. Expect to spend about 10 minutes.

1. Open **Filling Gaps.gsp** and go to page "Find Values." Enlarge the document window so it fills most of the screen.

2. Explain, *Here is a table of values that is partially complete, but has some missing values. Can you fill in the gaps?* Let students suggest ideas for how they might find the missing values. Encourage different approaches, but don't pursue them in detail.

3. After a few minutes, present the strategy that is used in this activity, whether or not this approach is suggested. *One method for filling in the gaps is to graph the pairs that you know and look for an equation that goes through them.* Model the point and function plotting in worksheet steps 2 and 3. Here are some tips.

   - Plot the point (4, 11). *Where is the point I just plotted?* Elicit the idea that the scale of the axes needs be changed, and then show how you can drag the control point at (1, 0) to change the scale.

   - Plot the three given points. *What might the graph that passes through these points look like?* Students will most likely suggest a line. *What would be the equation of that line?*

   - Finding the correct function in worksheet step 3 will depend on the background of your students. In most cases, students can find the correct equation using guess-and-check. If necessary, construct a line through the given points and choose **Measure | Equation.**

DEVELOP

Expect students at computers to spend about 25 minutes.

4. Assign student pairs to computers and tell them where to locate **Filling Gaps.gsp.** Distribute the worksheet. Tell students to work through step 10 and do the Explore More if they have time.

5. Let pairs work at their own pace. As you circulate, here are some things to notice.

   - If students working on worksheet step 2 don't see the **Plot as (x, y)** choice in the Graph menu, an object (perhaps the previously plotted point) is selected along with two numbers in the table. Students can either click on the extra object(s) to deselect, or they can click on blank space and then reselect the two coordinates in the table.

- Students will have to drag the control point $(1, 0)$ or the graph's scale to see all of the plotted points.

- Some students' points may not lie in a straight line. They may have chosen the coordinates to plot in different orders—sometimes from the left column first, sometimes from the right column first. Suggest repeatedly choosing **Edit | Undo Construct Plotted Point** and ask, *Which column should you choose first each time?*

- Worksheet step 4 encourages careful thinking about how the table and expression relate. You may need to ask some questions. *Where might you find that 3 in the graph? Where is 3 in the table? Where is −1 in the graph? What row in the table would correspond to that point on the graph?*

- On page "Find More Values," only two complete pairs of points are given. Some students may be struggling to find a linear expression to plot. *Have you looked back at page "Find Values"? Would it help you to plot ordered pairs and a function expression?*

- The tables in worksheet steps 8 and 9 are generated with random integers. For your information, the starting value (in the left column) is between −5 and 10, inclusive. In the right column, the common difference is between −5 and 5, and the "start value" (corresponding to 0 in the left column) is between −10 and 10. If you need to reset the values to the original (common difference 3 and start value 9), you can scroll down to find a Reset button.

- Direct students who finish early to go on to Explore More.

SUMMARIZE

Expect to spend about 10 minutes.

6. Call students together after worksheet step 10 or as needed to finish the class on time. Open **Filling Gaps.gsp** for reference. Ask for volunteers to describe their methods. At this point, they can assume that the common difference in the left column is 1. (In the Explore More, the common difference in the left column is not 1.)

7. *Why is the common difference in the right column being multiplied by x in the function expression?* Elicit the idea that repeatedly adding the common difference is the same as multiplying by the common difference.

8. If students have seen the notion of function, even briefly, you might build on that knowledge. ***Why does Sketchpad use the word* Function *in the Graph menu command* Plot New Function?** Encourage a variety of suggestions. Students may think of a function as a table of ordered pairs or as a rule for relating the left column to the right column. ***What's a function expression?*** [An algebraic way of describing that rule.] Alternatively, students may have encountered functions only in the context of function machines. ***How does the function expression here relate to a function machine?*** [It gives the rule governing the machine.]

9. ***What are functions or expressions in this form called?*** [They are linear, because the graphs are straight lines.]

10. If time allows, give students a chance to try some of the methods others have described by presenting challenges from pages "Finding More Values" and "Explore More."

11. ***What other questions can you ask that might be explored?*** Encourage a variety of responses such as these.

    What would happen if you entered an equivalent form of the expression?

    Why is the graph of a linear function a straight line?

    How many linear functions are there?

    Are there nonlinear functions?

    Can linear functions have other forms?

EXTEND

You might further challenge students to make tables to challenge each other. Suggest that they include negative constant differences in the right column.

ANSWERS

2. A (straight) line will pass through the points.

3. Function expressions are equivalent to $3x - 1$. The corresponding equation might be written $y = 3x - 1$.

4. Answers will vary. Sample answer: Find the common difference between values in the second column of rows 3, 4, and 5. Call that value m. Then work backward to determine what the second-column value would

Exploring Expressions and Equations in Grades 6–8 with The Geometer's Sketchpad
© 2012 Key Curriculum Press

 ACTIVITY NOTES

be in the row with first-column value 0. That's value *a*. The function's expression is then $mx + a$.

6. Coordinate pairs are $(2, 5)$, $(3, 8)$, $(4, 11)$, $(5, 14)$, $(6, 17)$, $(7, 20)$, and $(8, 23)$.

7. The given pairs are $(1, 12)$ and $(4, 39)$.

8. $y = 9x + 3$

9.–10. Answers will vary, because the table is generated with some random values. Pairs that students list should satisfy their equations.

11. Answers will vary, but they should extend the answer in worksheet step 4 to divide the common difference in the second column by the common difference in the first column. Answers could also use the language of in-out tables or include reference to finding a number that is the ratio of the change in the *y*-values to the change in the *x*-values.

Sample answer: Divide the common difference in the second column by the common difference in the first column. Call that value *a*. Then determine what the second-column value would be in the row with first-column value 0. Call that value *b*. The function's equation is $y = ax + b$.

Filling Gaps

 Name:

In this activity you'll investigate how a graph can help you fill in gaps in a table of values.

EXPLORE

1. Open the sketch **Filling Gaps.gsp.** On page "Find Values," you see a table with some values missing. One method for filling in gaps is to graph the pairs you know.

2. Choose **Graph | Show Grid.** Plot each of the three known points by selecting the *x*-coordinate, then the *y*-coordinate, and choosing **Graph | Plot as (x, y).** Adjust the scale to see the points.

3. Look for a graph that goes through the points. Choose **Graph | Plot New Function** and experiment with the expression until you get the graph to pass through all three points. Double-click the function to change it.

   What expression did you enter?

4. How could you find the equation directly from the table?

5. Construct a point on the line you plotted.

6. Choose **Measure | Coordinates.** Drag the point along the line until you find the missing values. Fill the gaps in the table. Then list the pairs of values from each row of the table as the coordinates of a point.

2	
3	
4	11
	14
6	17
8	23

Exploring Expressions and Equations in Grades 6–8 with The Geometer's Sketchpad
© 2012 Key Curriculum Press

7. Go to page "Find More Values." Here is a new table.

    List the pairs of values in the table as the coordinates of points.

8. What's the equation for the line through the points?

9. Press *New Challenge* and repeat steps 7 and 8.

    List the pairs of values in the table as coordinates of points.

    What is the equation for the line through the points?

10. Press *New Challenge* once more and repeat steps 7 and 8.

    List the pairs of values in the table as coordinates of points.

    What is the equation for the line through the points?

EXPLORE MORE

11. Go to page "Explore More." Here's another table with gaps. This time, the values in the first column do not differ by 1. How can you extend your equation-finding method to a method that works for tables like this?

    What pairs of numbers are missing from the table?

    What is the equation?

    Press *New Challenge* to check your method for other tables.

Mellow Yellow: Interpreting Graphs

 ACTIVITY NOTES

INTRODUCE

Project the sketch for viewing by the class. Expect to spend about 5 minutes.

1. Open **Mellow Yellow.gsp** and go to page "Story 1."

2. Explain, *Today you're going to put graphs into motion.* Pointing to Mellow Yellow, explain, *This is Mellow Yellow and she will be walking along this line segment, which represents the path from her house to the corner store. Of course, Mellow Yellow will not always walk in the same way. Sometimes she stops to pick something up or to rest. Sometimes she runs and sometimes she dillydallies.* Press *Go! Can anyone describe what Mellow Yellow did?* Let students talk about the different things they noticed. Encourage them to talk especially about the different speeds at which Mellow Yellow moved. *The graph provides a very nice way of describing Mellow Yellow's walk, and today you're going to work on being able to interpret the story a graph can tell.*

3. Show students how the points on the graph can be dragged to different positions and that the resulting walk done by Mellow Yellow will change. Try to avoid giving any of the segments a negative slope for now. Tell students that they will have an opportunity to change the graphs when they get to "Story 3," but that they should not change them for the first two stories.

DEVELOP

Expect students at computers to spend about 30 minutes.

4. Assign students to computers and tell them where to locate **Mellow Yellow.gsp.** Distribute the worksheet. Tell students to work through step 5 and do the Explore More question if they have time.

5. Let pairs work at their own pace. As you circulate, here are some things to notice.

   - Students often think that a slower speed means the graph goes down. Make sure students articulate what they see happening to Mellow Yellow's journey when the graph goes down, and how they can compare slower speeds to faster ones.

   - For worksheet step 3, where students are asked whether the graph corresponds to the story, invite students to think about how they could change the graph so that the story *does* match.

   - For worksheet step 4, encourage students to experiment with different locations of points 1 and 2 if they are having difficulty understanding the relationship of the graph to the movement of Mellow Yellow.

Invite students to change the locations of points 1 and 2, and describe a different story.

- For worksheet step 5, make sure students write out their stories fully. Ask them to connect each "leg" of the trip to the corresponding segment.

SUMMARIZE

Project the sketch. Expect to spend about 10 minutes.

6. Gather the class. Students should have their worksheets with them. Using page "Fit the Story," ask students to explain how they used the information in the story to know where to place the points on the graph.

7. Using page "Write a Story," ask two or three volunteers to describe their stories.

8. Ask students what kinds of motions they were able to create with the graph and also what kind of motions they could not create (such as acceleration).

EXTEND

1. Explain to students that acceleration and deceleration require nonlinear graphs. Point out that one of the problems with the graphs they used in Sketchpad is that Mellow Yellow would have a hard time starting immediately at a high speed. Instead she would probably start more slowly and accelerate until she achieved a high speed. On the board, draw an example of acceleration and ask students to use these curved lines to draw graphs that would better represent the stories about Mellow Yellow.

2. *What other questions might you ask about graphing motion?* Encourage all inquiry. Here are some ideas students might suggest.

   Why do steeper lines indicate faster speeds?

   Can you calculate the actual speed from the steepness of the line?

   What if a graphed segment of the trip were vertical?

ANSWERS

1. Answers will vary. Students should notice that the *x*-axis corresponds to the time it takes Mellow Yellow to travel and that the *y*-axis corresponds to the distance she has traveled. The *y*-coordinate of point *Stop* is just under 1 (mile) and its *x*-coordinate is just over 11, so it takes her just over 11 minutes to arrive at the corner store. There are three different slopes, including one that is 0 (corresponding to the horizontal segment) when Mellow Yellow is stopped.

2. In Story 2 Mellow Yellow goes backward and the corresponding segment slopes down.

3. The first segment should have a steeper slope than the last segment because she runs faster at the beginning. The middle segment should have a slope of 0 instead of a negative slope.

4. Answers will vary. Make sure that the first segment has a small positive slope, the second segment has a slope of 0, and the third segment is steeper than the first. Make sure also that Mellow Yellow runs far enough, so that the *y*-coordinate of *Stop* is just under 1.

5. Answers will vary. Make sure that the horizontal segments correspond to not moving and that the slopes of the other segments correspond appropriately to the speeds.

6. Answers will vary. Make sure that the *y*-coordinate of point *Stop* in both stories is just under 0.5 (miles) and that its *x*-coordinate is at 10 or 5 (minutes), depending on the story.

Exploring Expressions and Equations in Grades 6–8 with The Geometer's Sketchpad
© 2012 Key Curriculum Press

Mellow Yellow

 Name:

In this activity you'll try to describe and predict how different motions, such as stopping, walking slowly, or walking very quickly, are represented on a graph.

EXPLORE

1. Open **Mellow Yellow.gsp** and go to page "Story 1." Press *Go!*, and then press *Show Story.* Describe how the features of the graph (the axes, slopes, and points) correspond to the story of Mellow Yellow's walk.

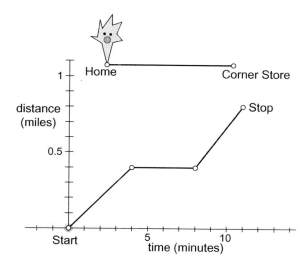

2. Go to page "Story 2." Read the story and compare the graph to the story. Then press *Go!*, and describe the different types of motion you see in Story 2 compared to Story 1.

3. Go to page "Story 3." Read the story and then press *Go!*. Decide whether the graph corresponds to the story. If not, change the graph (or the story!). Describe what you did.

4. Go to page "Fit the Story." Read the story and drag points 1 and 2 to make the graph fit the story. Check your graph by pressing *Go!*, and describe what you did.

5. Go to page "Write a Story." Write a story that fits the graph. Check your story by pressing *Go!*. Change your story if necessary. Then write your final story here.

EXPLORE MORE

6. Write your own story, but this time imagine that Mellow Yellow has to travel only to the bus stop, which is halfway to the corner store, in about 10 minutes. Then write a story in which she travels to the bus stop, but in about 5 minutes. Use pages "Explore More A" and "Explore More B" to fit the graph to each of your stories.

Exploring Expressions and Equations in Grades 6–8 with The Geometer's Sketchpad

Weather Balloon: Comparing Distance and Speed

INTRODUCE

Project the sketch for viewing by the class. Expect to spend about 5 minutes.

1. **Build on students' knowledge.** *Do you know what weather balloons do?* Some students may say that they predict the weather. Actually, they carry instruments to report on conditions of the upper atmosphere. These reports help with studying the upper atmosphere as well as with predicting the weather. *How high do you think they go?* Students may be surprised that weather balloons go up to 30 kilometers or more (more than 18 miles), much higher than large long-distance airplanes fly (10–11 km).

2. Pose the problem: *Your goal is to examine some data for a balloon's trip and predict how long it'll take to reach a height of 30 km. Then you'll compare different graphs that give information about the balloon as it rises.*

DEVELOP

Expect students at computers to spend about 25 minutes.

3. Assign pairs to computers and tell them where to locate **Weather Balloon.gsp.** Distribute the worksheet. Encourage students to ask their neighbors for help if they are having difficulty with Sketchpad.

4. Let pairs work at their own pace. As you circulate, here are some things to notice.

   - As needed, encourage students to make observations for quite a few balloon "flights." Help them observe both the table and the graph. *What's happening in the table? What's happening in the graph? Does the table relate to the graph? How?*

   - Worksheet step 5 is to help students understand the graph. They may believe that the plotted points are on the path of the balloon. If they're struggling with this question, ask leading questions to help them recognize that the horizontal axis represents time. Ask them to connect values in the table to the graph and describe what those values represent. *So what does it mean that the graph goes to the right from point to point?* [It shows that time has passed.]

   - In worksheet step 6, students might use different approaches. For later sharing, note students whose approaches vary.

   - In worksheet step 8, students can do these calculations using **Number | Calculate** and clicking on values in the sketch.

- In step 10, if students are stuck, offer the hint that it has something to do with the table they just completed in step 8.

SUMMARIZE

Expect to spend about 15 minutes.

5. Convene the class for sharing. Students should have their worksheets with them. Ask students to present their various approaches to step 6—finding when the balloon will reach 30 km. Answers through various approaches may or may not differ. If they do differ, ask, **Why do we get different answers?** Bring out the idea that round-off errors accumulate.

6. Open **Weather Balloon.gsp** and go to page "Simulation." Ask students to describe how the graphs are related. Bring out the idea that the speed is the slope of the time-height graph.

7. **What have you learned?** Try to elicit from students these lesson objectives.

   - The arrangement of points representing a moving object does not show the path of that object.

   - The average speed of an object is the distance traveled divided by the amount of time elapsed.

   - A time-distance graph for an object traveling at a constant speed is a straight line.

   - Speed is different from distance. Speed is the slope of the time-distance graph.

   - A time-speed graph for an object traveling at a constant speed is a horizontal line. A horizontal line represents a situation where the values of the second coordinates don't change.

8. If time permits, discuss the Explore More. You might follow up with the question, **What would happen to the graphs if we could change the speed to a negative value? What would that mean?** Negative speeds can be interpreted as the balloon going downward toward the earth, and negative heights can be interpreted as the distance below ground.

9. Probe for further questions. ***What other questions can you ask?***
Students may have a variety of ideas, including these.

Does the balloon keep going forever?

How high does it get?

What does the graph look like if the balloon falls?

Is a graph of heights at various times ever the same as a graph of the path of the object?

What are weather balloons made of?

Will the balloon really have a constant speed?

EXTEND

You might mention that a rubber balloon's ascent rate is fairly constant and suggest that interested students investigate why that is the case. Students might be interested in doing a little research on how weather balloons work. A balloon ultimately pops when it reaches too high an elevation and instruments attached to it parachute back to earth. You could encourage students to make time-height and time-speed graphs for these instruments.

ANSWERS

2. The points on the graph rise going from left to right and seem to lie in a straight line. They are evenly spaced.

3. The time values in the table are increasing by 20 minutes, and the height values in the table are increasing by 6.9 or 7 kilometers each time.

4. The points are in a line because going from one to the next is going over and up by the same amount each time (20 seconds and 6.9 km).

5. The first coordinate of the plotted points is time, not a horizontal distance. So the direction of the graph going to the right represents the passage of time, not the path of the balloon.

6. Answers may vary near 86 minutes. Justifications may vary. For example, students might construct a single line through all the points, construct another point on that line, measure its coordinates, and move it until the second coordinate is 30. Or they might repeatedly add 0.35 km/min until the sum reaches 30. Or they might divide 30 by 0.35.

Equivalently, students who prefer equations might think in terms of solving the equation $0.35x = 30$. Other students may work with 20-minute increments, dividing 30 by 6.9 to get about 4.35 increments, and then multiplying by 20 minutes per increment.

7. The point (86, 30) will lie on the line containing the other points representing the balloon's time and height. If the answer were incorrect, the point would not fall on the line.

8. Each average speed is 0.35 km/min.

10. The vertical axis on this graph represents speed.

11. The line is horizontal because the speed is constant—it doesn't increase or decrease.

12. Both graphs are straight lines, but the height graph is diagonal whereas the speed graph is horizontal.

13. As the rate increases, the height values in the table increase, and the graph becomes steeper. The rate of change in the graph (the slope of the line) is the speed.

15.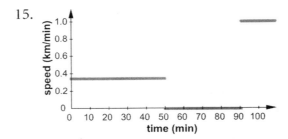

16.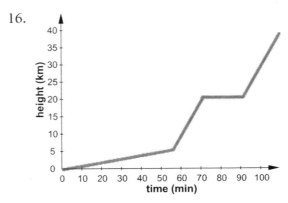

Weather Balloon

 Name:

In this activity you'll explore different graphs of a rising weather balloon and predict how long it will take the balloon to reach a height of 30 kilometers.

EXPLORE

1. Open **Weather Balloon.gsp** and go to page "Balloon." Press *Launch* and watch how the table and graph change as the balloon rises.

2. What do you observe about the points on the graph?

3. What do you observe about the table values?

4. Why do you think the points on the graph lie on a straight line?

5. The points on the graph represent the height of the balloon at different times. The balloon is rising vertically, but the points are not along a vertical line. Why not?

6. How long will it take the balloon to reach a height of 30 kilometers? Describe your reasoning.

7. Choose **Graph | Plot Points.** Enter the time you found as the first coordinate, and enter 30 as the second coordinate. Click **Plot**, and then click **Done.** How can you tell from this point that your answer is correct?

8. Fill in the third column in this table. To find the speed at each point in time, divide *Height* by *Time*. Round to the nearest hundredth. What do you notice?

Time (minutes)	Height (kilometers)	Speed (km/min)
20	6.9	
40	13.9	
60	20.8	
80	27.8	
100	34.7	

9. Go to page "Mystery Graph" and press *Launch.* The balloon travels in exactly the same way as it did on page "Balloon," but the graph is different—the vertical axis does not represent the balloon's height this time.

10. What does the vertical axis on this graph represent?

11. Why do the points on this graph lie on a horizontal line?

12. Go to page "Simulation" and press *Launch.* Then press *New Balloon, Show Traces,* and then *Launch* again. Describe how the top and bottom graphs are related.

13. Experiment with changing the speed and launching new balloons. When you increase the speed, what happens to the table? What happens to the graph? Explain why.

Exploring Expressions and Equations in Grades 6–8 with The Geometer's Sketchpad
© 2012 Key Curriculum Press

EXPLORE MORE

14. Go to page "Variable Speed." Now you can change the speed while the balloon is rising. Experiment with changing the speed as the balloon rises.

15. Sketch the speed graph related to the given height graph.

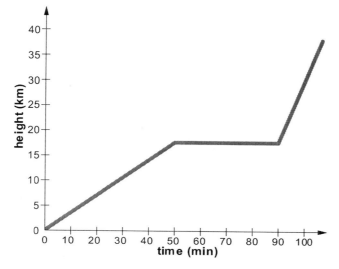

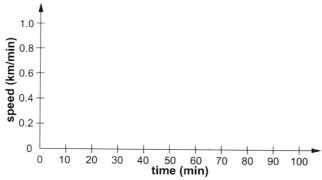

16. Sketch the height graph related to the given speed graph.

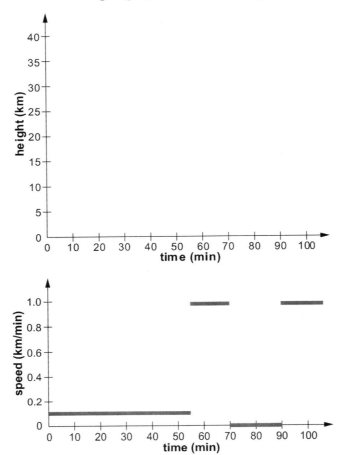

Exploring Expressions and Equations in Grades 6–8 with The Geometer's Sketchpad
© 2012 Key Curriculum Press

Dance Pledges: Plotting Linear Equations  ACTIVITY NOTES

INTRODUCE

Project the sketch for viewing by the class. Expect to spend about 10 minutes.

1. Open **Dance Pledges Present.gsp,** and go to page "Problem." Enlarge the document window so it fills most of the screen. Ask for volunteers to read the problem aloud. You might have students act it out.

2. *In a few minutes, you're going to work in pairs to compare the three ideas for getting pledges. But first, does their reasoning make sense?* Allow students to share their opinions and discuss what they think is realistic.

3. Go to page "Plots" and lead students through plotting a function for Chris's scheme. *How would you calculate how much Chris's scheme earns if Chris dances for 6 hours?* [$9] *8 hours?* [$12] *24 hours?* [$36] *What calculations are you doing?* [$1.50 times the number of hours] Choose **Graph | Plot New Function** and enter the expression $1.5 * x$.

4. Demonstrate how to construct a point on the graph by selecting the graph and choosing **Construct | Point on Function Plot** or by using the **Point** tool. Then select the point and choose **Measure | Coordinates.** Help students focus on the meaning of the coordinates. *Can we use the coordinates to find out how many hours Chris would have to dance to earn $18?* [Drag the point until the y-coordinate is 18 and find the x-coordinate. He would have to dance 12 hours.]

DEVELOP

Expect students at computers to spend about 25 minutes.

5. Assign student pairs to computers and tell them where to locate **Dance Pledges.gsp.** Distribute the worksheet. Tell students to work through step 12 and do the Explore More if they have time.

6. Let pairs work at their own pace. As you circulate, here are some things to notice.

   • Allow plenty of opportunity for students to help each other. Use questions similar to those you asked in the introduction: *How would you calculate how much Jesse's* (or *Dale's*) *scheme earns for 6 hours?* [$10 (or $6)] *8 hours?* [$10 (or $7)] *What calculations are you doing?*

   • You might need to show students how to adjust the scales on the axes to see their plots.

- If students don't see the **Plot New Function** choice in the Graph menu, they have some other object selected. They can click in blank space to deselect all objects.

- Jesse's scheme might be challenging because students may believe that the function expression must contain an *x*.

- In Dale's plan, students might use 50 instead of 0.5 for the hourly amount. *What measurement units are you using?*

- You might suggest that students label the graphs so that they know which represents whose plan.

SUMMARIZE

Continue to project the sketch. Expect to spend about 10 minutes.

7. Have a few selected pairs share their insights. Introduce or review the terms *slope* and *y-intercept*. *Can you tell what the slope and y-intercept are from the equation?* [Yes] *If you know that the slope was 3 and the y-intercept was 7, what would the equation be?* [$y = 3x + 7$]

8. Encourage discussion of the question in worksheet step 9. The answer depends on the assumptions made about dancing parts of hours. Students may argue that the payments are only by the hour, so Chris's scheme never pays exactly $10. Depending on the background of your students, you can also have them use substitution to find the points of intersection algebraically.

9. *What have you learned?* Bring out these objectives.

   - The graph of $y = $ (hourly amount) $x + $ (fixed amount) is a straight line.

   - The amount multiplied by *x* is the slope of the line, measuring its steepness.

   - The amount added is the *y*-intercept of the line, measuring its height along the *y*-axis.

   - The graph of $y = $ (fixed amount) is a horizontal line.

   - A point of intersection of two graphs can show where two different plans earn the same amount of money for the same amount of work.

 ACTIVITY NOTES

10. ***What other questions can you ask?*** Some interesting questions
are these.

*What if the slope or y-intercept were negative? What would the equation
and graph look like?*

Are there any lines that can't be plotted?

ANSWERS

1. $y = 1.5x$

2. $30

3. $y = 10$

4. Answers will vary. Both graphs are straight lines. Jesse's graph is
horizontal but Chris's is not. (It rises from left to right.)

5. $10

6. $y = 5 + 0.5x$ or $y = 0.5x + 5$

7. The earnings for dancing 0 hours.

8. $15

9. Jesse's scheme always brings in $10 per donor. Dale's scheme brings
in $10 for dancing 10 hours. Chris's scheme brings in $10 for dancing
$6\frac{2}{3}$ hours.

10. Points representing the same amount of money for the same number
of hours are at the intersections of the graphs. Chris's and Dale's plans
both earn $7.50 for 5 hours of dancing. Dale's and Jesse's plans both
earn $10 for 10 hours of dancing.

11. As the hourly amount increases, the line gets steeper, or the graph rises
more for each hour danced.

12. As the fixed amount increases, the height of the line as it passes
through the *y*-axis increases.

13. Answers will vary.

Dance Pledges

 Name:

Three friends are planning to be in a dance marathon to raise money for kids with cancer.

Chris says, "I'm going to dance for the whole 24 hours, but nobody thinks I'll make it more than 2 or 3 hours, so I'll ask them to pledge $1.50 an hour."

Jesse says, "I know I can't go for more than a few hours, so I'm going to ask for $10 no matter what."

Dale says, "I have no idea how long I can go. I think I'll ask for $5 plus 50 cents an hour."

 1. Open the document **Dance Pledges.gsp** and go to page "Plots."

   Plot a function for Chris's plan, using **Graph | Plot New Function.** What equation did you use?

 2. Construct a point on the graph. Select the point and choose **Measure | Coordinates.**

   Drag the point. How much will each donor pay if Chris dances 20 hours?

3. Plot a function for Jesse's plan. What equation did you use?

4. How does this graph compare to Chris's?

 5. Construct a point on Jesse's graph and measure its coordinates.

   Drag the point. How much will each donor pay if Jesse dances 20 hours?

6. Plot a function for Dale's plan. What equation did you use?

7. What does the point where the graph intersects the *y*-axis represent?

Exploring Expressions and Equations in Grades 6–8 with The Geometer's Sketchpad
© 2012 Key Curriculum Press

8. Construct a point on the graph and measure its coordinates.

Drag the point. How much will each donor pay if Dale dances 20 hours?

9. For how much dancing time does each plan bring in $10 per donor? Explain.

10. Do any of the plans ever bring in the same amount of money for the same number of hours danced? Explain.

 11. Go to page "Sliders." You'll see sliders for fixed and hourly amounts in a pledge plan. Choose **Graph | Plot New Function**. Instead of entering numbers, click *Fixed Amount* and *Hourly Amount*. What happens to the graph as you change the *Hourly Amount* slider? Explain.

12. What happens to the graph as you change the *Fixed Amount* slider? Explain.

EXPLORE MORE

13. Create a scheme that does not produce a straight line when graphed.

Hikers: Solving Through Multiple Representations

INTRODUCE

Project the sketch for viewing by the class. Expect to spend about 10 minutes.

1. Open **Hikers Present.gsp** and go to page "Problem." Enlarge the document window so it fills most of the screen. Read the problem aloud. Pause when you come to the first occurrence of mi/h, and ask students how to read it and what it means.

2. You might have two volunteers act out the problem for the class. Then go to page "Simulation" and press *Start/Stop Simulation*. **Between what times do the hikers meet?** [Between 3 and 4 hours].

3. **What might we do to find the time more precisely?** Encourage many suggestions. Go to page "Table".

4. **Do the distances corresponding to time 0 make sense?** Help students see that the distances are from the trailhead, not the distances traveled, so Maria begins at 12 miles.

5. **What should go on the row corresponding to time 1?** Elicit the idea that Edna will be 1.5 miles from the trailhead and Maria will be $12 - 2 = 10$ miles from the trailhead. Demonstrate how to double-click the parameters to change their values.

DEVELOP

Expect students at computers to spend about 25 minutes.

6. Assign student pairs to computers and show them how to find **Hikers.gsp.** Distribute the worksheet. Ask students to work through step 12 and do the Explore More if they have time.

7. Let pairs work at their own pace. As you circulate, here are some things to notice.

   - Some students may say that the hikers meet 6 miles from the trailhead, because 6 appears in both columns (or because 6 is the halfway point). Ask, **How many hours after they left were they 6 miles from the trailhead? Can you say that's the same time? What does it mean to meet?**

   - Students may have difficulty reading the table to see when the hikers meet. **Who was closer to the trailhead after 2 hours? After 3 hours? After 4 hours?**

   - In worksheet step 3, if students don't see **Graph | Plot as (x, y),** they probably still have the previous point selected. Tell them to click in blank space before they plot each point.

- If the lines that students trace in worksheet steps 7 and 8 do not pass through the plotted points from step 3, tell them to change their expressions in step 6, erase their traces, and try again.

- The coloring in worksheet step 9 is intended to help students associate the graphs directly with the table.

- In worksheet step 10, students may not realize that the point they seek is the intersection. Help them think of one variable at a time. *What points represent the two hikers at a distance 9 miles from the trailhead? What are their times when they are there? What points represent the hikers 6 miles from the trailhead? What are their times? What point represents when they are at the same place at the same time?*

- As needed, help students realize in worksheet step 11 that to find exact times and distances, they must convert to fractions.

SUMMARIZE

Expect to spend about 10 minutes.

8. Reconvene the class. Select some pairs to present their sketches (or use those in **Hikers Present.gsp**). Discuss worksheet steps 5–11.

9. *Which approach do you prefer: a table, a graph, or an equation?* Student preferences will vary. Students should realize that numerical information can be represented in multiple ways: arithmetically, algebraically, and graphically. Encourage comparisons of the methods, such as the fact that the table and graph give only estimates, whereas the equation could give an exact answer.

10. *What have you learned?* You may wish to have students respond individually in writing to this prompt, or have volunteers respond verbally. Bring out these objectives.

   - The same situation can be represented with tables, graphs, or equations.

   - Making a table can help in finding expressions to graph.

   - Knowing expressions to graph can help in finding an equation to solve.

EXTEND

What other questions might we ask? Encourage all student curiosity. Mathematical questions of interest include these.

Why are the graphs straight lines?

Why does the point of intersection represent where the hikers met?

Is there an easier way to find an exact solution?

What if the hikers paused to rest? Could we still tell when they met?

Are there ways other than tables, graphs, and equations to represent the situation?

What if there were no Maria? If Edna went over one day and came back the next, would there necessarily be a point at which she was at the same place at the same time of day?

ANSWERS

1.

Time (hours)	Edna's Distance (miles from trailhead)	Maria's Distance (miles from trailhead)
0	0.0	12.0
1	1.5	10.0
2	3.0	8.0
3	4.5	6.0
4	6.0	4.0
5	7.5	2.0

2. The hikers will pass each other between hour 3 and hour 4. Any students who put in extra rows between hour 3 and hour 4 may predict a smaller interval.

3. The points representing each hiker's distances lie in a straight line, and those lines will cross, though not at a data point. The sketch should look like this.

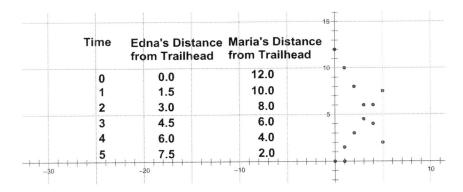

Exploring Expressions and Equations in Grades 6–8 with The Geometer's Sketchpad
© 2012 Key Curriculum Press

6. Edna: $1.5x$; Maria: $12 - 2x$

7. The traced point moves in a straight line, passing through the plotted points representing Edna's distance.

8. The traced lines intersect when *Time* is approximately 3.4 hours.

10. Answers may vary. The point of intersection will be approximately $(3.43, 5.14)$, indicating a time of 3.43 hours and a distance of 5.14 miles from the trailhead.

11. $1.5x = 12 - 2x$

   $3.5x = 12$

   $$x = \frac{12}{3.5} = \frac{12}{\frac{7}{2}} = \frac{24}{7} \text{ hours}$$

   $$1.5\left(\frac{24}{7}\right) = \frac{3}{2} \cdot \frac{24}{7} = \frac{36}{7} \text{ miles}$$

12. $\frac{24}{7}$ hours $= 3\frac{3}{7}$ hours

   $$= 3 \text{ hours} + \frac{3}{7} \cdot 60 \text{ minutes}$$

   $$= 3 \text{ hours and about } 25 \text{ minutes}$$

   Some students may go further to add about 43 seconds.

13. Answers will vary. Students familiar with solving systems of equations simultaneously may write two equations and solve by substitution (equivalent to what is done in worksheet step 8) or by elimination.

   $$\begin{aligned} 1.5x - y &= 0 \\ 2x + y &= 12 \\ \hline 3.5x &= 12 \end{aligned}$$

   $$x = \frac{12}{3.5} \approx 3.43 \text{ hours}$$

   $$y \approx 1.5(3.43) \approx 5.14 \text{ miles}$$

   Alternatively, students might think about closing speed. The hikers together need to cover 12 miles at a combined speed of $1.5 + 2 = 3.5$ miles per hour. Doing so will take $\frac{12}{3.5}$ hours.

   Solutions could also be estimated by finding the intersection of the graphs of equations $y = 3.5x$ and $y = 12$.

Hikers

Edna leaves a trailhead at dawn to hike toward a lake 12 miles away, where her friend Maria has been camping. At the same time, Maria leaves the lake to hike toward the trailhead (on the same trail, but in the opposite direction).

Edna is walking uphill, so her average speed is 1.5 mi/h. Maria is walking downhill, so her average speed is 2 mi/h.

In this activity you'll investigate when and where the hikers will meet.

EXPLORE

1. Complete the table.

Time (hours)	Edna's Distance (miles from trailhead)	Maria's Distance (miles from trailhead)
0	0	12
1		
2		
3		
4		
5		

2. Open **Hikers.gsp** and go to page "Table." Double-click each value and change it to match the value in the table above. From the table, what can you predict about when the hikers will meet?

To predict the meeting time more accurately, you'll graph the points in the table, and then graph lines through them.

3. To graph Maria's first point, select the 0 in the Time column, then the 12 in Maria's Distance column, and choose **Graph | Plot as (x, y).** Axes and a grid will appear, with the point (0, 12) plotted. Repeat this process to plot all of Edna's and Maria's distances in the table. What patterns do you see?

4. Construct a point on the *x*-axis. While the point is still selected, choose **Measure | Abscissa (x).**

5. Change the label of the point on the axis to *T* and the label of the abscissa measurement to *Time.*

Exploring Expressions and Equations in Grades 6–8 with The Geometer's Sketchpad
© 2012 Key Curriculum Press

6. Choose **Number | Calculate** and use the Calculator to enter an expression for each hiker's distance from the trailhead. To enter *Time* into the Calculator, click the value in the sketch. What expressions did you use?

   Edna:

   Maria:

7. Select *Time* and your value for Edna's distance from step 6. Choose **Graph | Plot as (x, y)** and then choose **Display | Trace Plotted Point.** Drag point *T* and describe what you see.

8. Repeat step 7 using *Time* and your value for Maria's distance. When will the two hikers meet?

9. Choose **Display | Erase Traces.** Choose **Graph | Plot New Function** and enter an expression for each hiker's distance from the trailhead after *x* hours. Color each graph with the same color as the column it represents.

10. Construct a point on Maria's graph. While it's selected, choose **Measure | Coordinates.** Drag this point to predict when and where the hikers will meet. What are coordinates of that point? What does each coordinate represent?

11. Use the expressions you graphed to write a single equation. Solve the equation for the time when they meet, and then find the distance from the trailhead. Show your work.

12. Convert your time solution to hours and minutes.

EXPLORE MORE

13. You've made predictions by a table, by a graph, and by solving an equation. What other ways can you use to predict when the two hikers will meet?

Car Wash: Using a Graph to Make Decisions

INTRODUCE

Project the sketch for viewing by the class. Expect to spend about 5 minutes.

1. Open **Car Wash.gsp** and go to page "Situation." Ask for a volunteer to read the situation aloud.

2. Have the class develop the problem. Ask, **What questions might you investigate about this situation?** Record student questions directly on the Sketchpad page or on some other display. Many questions such as these will arise.

   How much will you charge per car?

   Do you have any expenses?

   How much profit will you make?

   How many cars do you have to wash to earn enough money?

   How many people will do the washing?

   How many cars can you wash at a time?

   How long will it take to wash a car?

3. Decide as a class on some realistic assumptions to make in order to calculate the profit. Have students debate these questions.

   What's a good price to charge per car? How much do you think people would be willing to pay for a car wash?

   What are some realistic expenses? How much will you need to pay for space, water, soap, sponges, rags, buckets, signs, and so on? Will any of this be donated, perhaps by a local service station?

   How much profit do you need to accomplish the coach's goal? How many cars must be washed to reach this profit?

   At this time, you might choose to also discuss the time, space, and manpower constraints, which will limit the maximum number of cars that can be washed. You might also choose to save this discussion for later. In these notes, this discussion is presented in the Summarize section.

DEVELOP

Continue to project the sketch. Expect to spend about 30 minutes.

4. Distribute the worksheet. Give students enough time to discuss worksheet steps 1–4 with their groups and write down their answers. Some groups might assume that the expenses depend on the number of cars washed, which is fine for now.

Exploring Expressions and Equations in Grades 6–8 with The Geometer's Sketchpad
© 2012 Key Curriculum Press

5. Briefly reconvene the class. *How do you calculate the profit?* Make sure students understand that profit is income minus expenses, and that income is the price per car multiplied by the number of cars. Tell students to assume that the expenses are constant, since the team will have to buy supplies before they actually have the car wash. You might also choose to reach a class consensus on reasonable values for the price per car and for the expenses.

6. *Now you'll make a graph to help analyze the situation.*

   If students have access to computers:

   Assign pairs to computers and tell them where to locate **Car Wash.gsp.** Depending on your students' prior experience with Sketchpad, you might want to model the construction of the function plot in worksheet steps 6–10. Then tell them to work through step 20 and do the Explore More if they have time.

   If students do not have access to computers:

   Have students make a graph on paper, starting with the two coordinates from worksheet step 3, adding three additional coordinates, and then connecting them with a line. Tell students to use their graphs to answer the questions in worksheet steps 7–16, and to pick a point on their line for step 8. You'll discuss steps 17–20 as a class during the Summarize section.

7. Give students enough time to answer at least worksheet steps 13–16. As you circulate, let students know that you can adjust the axes by dragging the double arrow that appears when the pointer is over the scale number. Make sure students understand that the profit can be negative if very few cars are washed. You might ask, *What would the profit be if you only washed 5 cars?*

SUMMARIZE

Expect to spend about 25 minutes.

8. Reconvene the class. Students should have their worksheets with them. Open **Car Wash Present.gsp** and go to page "Coordinates."

9. *Let's say you charge 5 dollars per car and the expenses total 80 dollars.* Set the Price slider to 5.00 and the Expenses slider to 80.00. *How much profit would your team make if you washed 60 cars?* [$250] Drag the point on the function plot to (60, 250) and ask someone to explain how to read this information from the graph. Do the same for 100 cars.

10. *Can the number of cars be negative?* [no] Discuss how this is meaningless in this context, even though the graph extends into the third quadrant. *Can the profit be negative?* [yes] Students should see that when the point is in the fourth quadrant, the *y*-value is negative. *What does this mean?* [income is less than expenses] Students might say, *If we don't wash enough cars, we would make less than what we had to pay in expenses.*

11. *How many cars must be washed to break even?* [16] As needed, define *break even* as having no loss, but no profit, or a profit of 0. Bring out the idea that this point is represented by the *x*-intercept of the line. *Does that make sense?* Students should see that the calculation of income for this number of cars is equal to the expenses.

12. *How much profit do you need to accomplish the coach's goal?* [$1075] *What can we say about the coordinates when the profit is $1075?* [The *y*-coordinate will be 1075.] Drag the point to (225, 1075). You may need to adjust the scales on one or both axes. *How many cars must be washed to accomplish the coach's goal?* [225]

13. *Does this seem like too many cars?* While this depends on the values your class has chosen, you will likely end up with a number over 100, which is a lot of cars to wash in one day. *What's a realistic number of cars to wash?* To arrive at a realistic maximum number of cars that can be washed, ask questions such as, *How many cars can be washed at a time? How many students will work? How many students will be working on one car? How long will it take them to wash it? How long will the car wash be open?* Decide on reasonable values and perform the calculations as a class to determine the most cars the team could expect to wash.

14. *What changes will allow your team to wash fewer cars and still make a profit of $1075?* Some student responses might include, *Charge more for each car, Lower the expenses,* or *Keep the car wash open longer.* Take all suggestions, and then use them as a lead in to see what happens to the graph if you change the price per car or the expenses. Go to page "Assumptions."

15. *So let's try changing the price.* Adjust the price slider. *What's happening?* Students might say, *The line is changing direction. It's changing steepness. It's rotating around the y-intercept. Why?* Through discussion, elicit the realization that as the cost per car changes, the

amount the graph rises for each one-car increase changes accordingly. Review or introduce the term *slope*, and point out that the slope is the coefficient of *x* in the equation of the line. ***In this problem, the slope is the rate: price per car.*** You might write $\frac{\text{Price}}{\text{Car}}$ on the board.

16. ***Another suggestion was to cut expenses. What happens to the graph as we change expenses?*** Some students may say that the line shifts horizontally. ***Why might you say the line shifts horizontally?*** [It shifts as the break-even point changes. The fewer the expenses, the lower the break-even point.] Other students may say that the line shifts vertically. ***Why might you say the line shifts vertically?*** [The expenses determine the *y*-intercept of the line.] Point out that the *y*-intercept is the constant in the equation of the line.

17. Bring the problem to a close by asking, ***Do you think it's realistic to achieve the coach's goal?*** Depending on time, let students argue their opinion, as long as they can support it with values they think are reasonable. You might also have them do this for homework.

18. ***What other questions might you ask?*** Here are some that might arise.

    Will expenses increase if we have lots more cars?

    Could we charge more for larger cars than for smaller ones?

    Will higher prices discourage customers?

19. ***What have you learned from this experience?*** You may wish to have students respond individually in writing to this prompt, or you may try to bring out these objectives.

    • *Profit* = (*Price per item*)(*Number of items*) − *Expenses*

    • The break-even point is the number of items for which the profit is 0 (the total income equals the expenses). It is the *x*-intercept of the graph.

    • For fixed expenses and price per item, the graph of the profit function is a straight line.

    • The slope of that straight line represents the price per item.

    • The slope of a line is the coefficient of *x* in the line's equation.

    • The *y*-intercept of that line represents the expenses.

ACTIVITY NOTES

- The *y*-intercept of a line is the constant in the line's equation.

- A mathematical representation (model) of a situation is limited; it can inform decisions but can't make them.

ANSWERS

1. Answers will vary.

2. Answers will vary. Here, we assume the price per car is $5 and the total expenses are $80.

3. For 60 cars, the profit is $220; for 100 cars, the profit is $420.

4. *Profit = (Price per car)(Number of cars) − Expenses*

7. The function is a line. The *x*-coordinate represents the number of cars.

11. The horizontal line intersects the *y*-axis at the *y*-coordinate of point *P*. The vertical line intersects the *x*-axis at the *x*-coordinate of point *P*.

13. Answers will vary, but they should be the same as in step 3.

14. A negative number of cars is meaningless in this situation. A negative profit is possible for a very small number of cars, because the income would be less than the expenses.

15. Answers will vary. Using the assumptions in step 2, the team would need to wash 14 cars to "break even." This point is the *x*-intercept of the graph.

16. Answers will vary. Using the assumptions in step 2, the team would need to wash 231 cars to make a profit of $1075.

18. The slope of the line changes. Increasing the price per car makes the line more vertical; decreasing the price per car makes the line more horizontal.

19. The *y*-intercept of the line changes. Increasing the expenses will shift the line down; decreasing expenses will shift the line up.

20. Answers will vary.

Car Wash

Your basketball team is planning the annual car wash to raise money for the team. Your new coach hopes that you can raise $275 so your team can attend a new tournament and $800 for some new equipment. How many cars must be washed to accomplish the coach's goal?

DISCUSS

1. Describe how you might solve this problem.

2. Make some assumptions to solve this problem.

   Price per car = _____

   Total expenses = _____

3. Based on these values, how much profit would your team make if you washed 60 cars? What if you washed 100 cars?

4. Write an equation for calculating the profit based on the number of cars washed.

CONSTRUCT

5. Open **Car Wash.gsp** and go to page "Experiment."

6. You'll plot a graph of the profit as a function of the number of cars washed. Choose **Graph | Plot New Function**. Then click on the values for *Price* and *Expenses* in the sketch and use the calculator keys to enter the expression:

   Price ∗ *x* − *Expenses*

   Click **OK**, and then choose **Display | Line Style | Thick**.

7. Describe the shape of the graph. What does *x* represent?

8. Construct a point on the graph. Drag it to make sure it stays on the graph. Label the point *P*.

9. Select point *P* and choose **Measure | Coordinates**.

10. Now you'll construct horizontal and vertical lines through point *P*. First, select point *P* and the horizontal axis and choose **Construct | Perpendicular Line**. Make the line dashed.

    Then construct another dashed line through point *P* perpendicular to the vertical axis.

11. Explain how the dashed lines are related to the coordinates of point *P*.

EXPLORE

12. Adjust the *Price* and *Expenses* sliders to match the values that you picked in step 2.

    Drag point *P* and observe its coordinates. Use point *P* to answer the questions in steps 13–16.

13. How much profit would your team make if you washed 60 cars? What if you washed 100 cars? How do these values compare to those in step 3?

14. Is it possible for the number of cars to be negative? Is it possible for the profit to be negative? Explain.

15. How many cars must be washed just to pay for the expenses? Where is this point located on the graph?

16. How many cars must be washed to accomplish the coach's goal? Is it possible for your team to wash that many?

Now you'll explore what happens to the profit graph if you change the price per car or the expenses. This might help your team make some decisions.

17. Delete the dashed lines through the moveable point on the graph.

18. Drag the *Price* slider. What happens to the graph? Explain why.

19. Drag the *Expenses* slider. What happens to the graph? Explain why.

20. Do you think it's realistic to achieve the coach's goal? Explain.

EXPLORE MORE

21. Make up a new situation involving price, expenses, and profit that you could model using this sketch.

Nonlinear Functions

Predicting Points: Linear and Quadratic Graphs

Students drag points, substitute coordinates, and use patterns to predict the graphs of equations, and then check by plotting the graphs. They start with linear equations and move to quadratic equations.

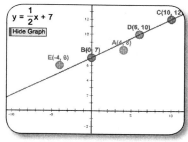

Grab Bag Graphs: Composing Symbols into Graphs

Students are given a "grab bag" of symbols, including variables, constants, and operations. They combine these to create linear and quadratic functions. Students use Sketchpad to plot their grab bag functions and to explore the changes that occur in the graphs when one item from the grab bag is removed from or added to a function.

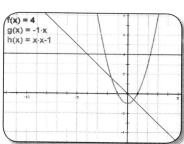

Million Dollar Fortune: Comparing Rates of Growth

Students choose among four different fortune cookies containing different options for receiving money. They investigate the four options, using tables and graphs to decide which option is best for different periods of time.

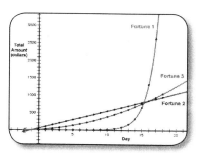

Match It Up: Comparing Graphs, Tables, and Equations

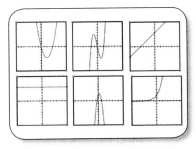

Students practice matching a given linear, quadratic, or exponential equation to one of six graphs or tables. They explore distinguishing features of each type of function and are able to compare and contrast linear, quadratic, and exponential relationships.

Overlapping Squares: Area and Perimeter Functions

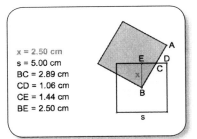

Students investigate the area and perimeter of the region formed by two overlapping squares. By moving one of the squares and observing the change in the overlapping area and perimeter, students determine the nature of a constant function and a quadratic function in a geometric setting.

Dancing Dynagraphs: Exploring Quadratic Functions

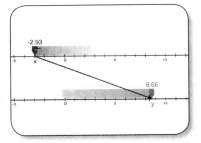

Students investigate properties and behaviors of quadratic equations using dynagraphs (graphs in which the *x*-axis and the *y*-axis are parallel). The quadratic equations are expressed in standard form. They compare these properties and relationships to linear equations.

Exploring Expressions and Equations in Grades 6–8 with The Geometer's Sketchpad
© 2012 Key Curriculum Press

Predicting Points:
Linear and Quadratic Graphs

 **ACTIVITY NOTES**

INTRODUCE

Project the sketch for viewing by the class. Expect to spend about 5 minutes.

1. Open **Predicting Points.gsp** and go to page "Linear." Have a volunteer read the instructions aloud. Ask, **What are we supposed to do?** Elicit the idea that we are to predict the graph of the equation and move the five points until they lie on that graph.

2. **Where might we move point A?** Students might suggest, *Move A to* $(1, 0)$, or, *Move A three units to the right and 5 units down.* You need not judge students' suggestions at this point. Review the term *coordinates* and what it means. **How can we tell whether a point is exactly where we want it?** [Check the point's coordinates.]

3. When the class has decided where to place all of the points, press *Show Graph.* Depending on how well the class did, you might discuss why some points did not end up on the line.

4. **How can we anticipate whether a point will lie on the graph?** Bring out the idea that its coordinates satisfy the equation—that is, when the point's coordinates are substituted for x and y in the equation, the two sides are equal.

DEVELOP

Expect students at computers to spend about 30 minutes.

5. Assign students to computers and tell them where to locate **Predicting Points.gsp.** Distribute the worksheet. Ask them to work through step 10 and do the Explore More if they have time.

6. Let students work at their own pace. As you circulate, here are some things to notice.

   - Some students may have difficulty adjusting the points precisely. Suggest that they look at the coordinates as they move each point.

   - Encourage students to look for patterns. For example, with lines students may see that they can place points by moving to the right one unit and then up or down according to the coefficient of x, thus previewing *slope.* For quadratics, the amount of vertical change is not constant, but there's a pattern to its change.

   - The first quadratics page has no x term. **What is the line of symmetry for the parabola?** [The y-axis.]

- While working with quadratics, the coefficient of the x^2 term will sometimes be zero. This makes the equation linear and the graph a straight line.

- Students may need to rescale an axis to be able to place five points with integer coordinates on the graph. They can do so by moving the cursor over a number in the scale so that the cursor becomes a double arrow, then drag.

- If students work on Explore More, you might suggest that they make a table for recording differences between consecutive values, differences between those differences, and so on.

SUMMARIZE

Expect to spend about 10 minutes.

7. Reconvene the class. **What have you learned?** Bring out these objectives, introducing or reviewing terminology as appropriate.

   - A point lies on the graph of an equation if the point's coordinates satisfy the equation.

   - The graph of linear equation $y = ax + b$ is a line.

   - For a linear equation, if a point lies one unit to the right of another point, then the difference in their y-coordinates is the coefficient of the x.

   - Equations of the form $y = ax^2 + bx + c$ are quadratic.

   - The graph of a quadratic equation is a parabola.

   - If there is no x term, the parabola is symmetric on the y-axis.

   - For a quadratic equation, the amount of change in y-coordinates increases by twice the coefficient of the x^2 term.

8. **What other questions can you ask that you may or may not be able to answer?** Encourage student curiosity. Here are a few interesting mathematical questions.

   Is it always the case that graphs of quadratics will be parabolas?

   Can we predict when the parabola will open downward?

   Could we see patterns more easily if we made a table?

Explore More

9. Ask students who completed the Explore More to describe patterns they found in cubic equations.

ANSWERS

2–4. Answers will vary, depending on what equations Sketchpad generates and where students drag points. The coordinates of the points should satisfy the equations.

5. Descriptions will vary. Students might mention the starting value where $x = 0$, moving up and over according to the slope, and substituting numbers in the equation.

6–9. Answers will vary, depending on what equations Sketchpad generates and where students drag points. The coordinates of the points should satisfy the equations.

10. Descriptions will vary. Students might mention substituting numbers in the equation and looking at the coefficient of x^2 to determine what the differences of differences will be.

11. For quadratic equations, the differences of the differences was a constant, twice the coefficient of the x^2 term. For cubic equations, the third differences are constant, six times the coefficient of the x^3 term.

Predicting Points

 Name:

In this activity you'll predict where points should go so that they will be on the graph of an equation you are given.

EXPLORE

1. Open **Predicting Points.gsp** and go to page "Linear." You'll see an equation in blue.

2. Drag the five colored points until you think they'll lie on the graph of the equation. Record the equation and the final coordinates of the five points.

   Equation:

   A: (_____, _____) B: (_____, _____) C: (_____, _____) D: (_____, _____) E: (_____, _____)

3. Press *Show Graph.* How many of your five points lie on the graph?

4. Press *New Equation.* Repeat steps 2 and 3 three more times.

   New equation:

   A: B: C: D: E:

   How many of your five points lie on the graph?

   New equation:

   A: B: C: D: E:

   How many of your five points lie on the graph?

   New equation:

   A: B: C: D: E:

   How many of your five points lie on the graph?

5. Describe your method for determining the location of the five points on the line. What patterns did you use?

6. Go to page "Quadratic 1." These graphs will be parabolas. Record the equation and list the coordinates of the five points on the equation.

   Equation:

   A: B: C: D: E:

   How many of your five points lie on the graph?

Exploring Expressions and Equations in Grades 6–8 with The Geometer's Sketchpad
© 2012 Key Curriculum Press

7. Press *New Equation.* Repeat step 6 two more times.

    New equation:

    A: B: C: D: E:

    How many of your five points lie on the graph?

    New equation:

    A: B: C: D: E:

    How many of your five points lie on the graph?

8. Go to page "Quadratic 2." This graph will also be a parabola. Record the equation and list the coordinates of the five points on the equation.

    Equation:

    A: B: C: D: E:

    How many of your five points lie on the graph?

9. Press *New Equation.* Repeat step 8 two more times.

    New equation:

    A: B: C: D: E:

    How many of your five points lie on the graph?

    New equation:

    A: B: C: D: E:

    How many of your five points lie on the graph?

10. Describe your method of locating the five points on the parabola. What patterns did you use?

EXPLORE MORE

11. Go to page "Explore More." Look for patterns in cubic equations, which have the form $y = ax^3 + bx^2 + cx + d$.

Grab Bag Graphs:
Composing Symbols into Graphs

INTRODUCE

Project the sketch for viewing by the class. Expect to spend about 5 minutes.

1. Open Sketchpad and enlarge the document window so it fills most of the screen.

2. Explain, *Today you're going to look more closely at the different symbolic components that go into algebraic equations. Some you've known for a long time, like the addition or multiplication sign. In this activity you'll play with a "grab bag" of symbols to see what different kinds of graphs you can create. You'll be seeing linear and quadratic equations of all kinds. Before you begin, I'll demonstrate how to create your graphs in Sketchpad.*

3. Using the **Text** tool, write down the following symbols in the sketch: $2, x, +$. Explain, *With just these symbols, I can create the expression $x + 2$. And I could use it to graph the function $y = x + 2$.* Choose **Graph | Plot New Function**, enter $x + 2$ in the dialog box, and press **OK**. Then ask, *What other functions could I make using those symbols?* Students should find $y = 2$, $y = x$, or $y = 2x$. Show students how to edit a function by double-clicking it with the **Arrow** tool. Ask the students to describe the three functions they have created. Explain, *In this activity you'll work with a different grab bag to create more interesting graphs.*

4. If you want students to save their work, demonstrate choosing **File | Save As,** and let them know how to name and where to save their files.

DEVELOP

Expect students at computers to spend about 20 minutes.

5. Assign students to computers. Distribute the worksheet. Ask students to work through step 6 and do the Explore More if they have time.

6. Let pairs work at their own pace. As you circulate, here are some things to notice.

   • Students should use each symbol only once in a given function.

   • Encourage students to use the preview window in the Calculator (the area just above the box where expressions are entered) when creating their functions.

   • Help students think about whether the graphs they see are quadratic, linear, or constant.

- In worksheet step 4, students might have trouble interpreting the graphs that do not have any scaling information on them. They can assume that both axes have the same scale, but otherwise encourage them to work with plausible numbers.

SUMMARIZE

Project the sketch. Expect to spend about 5 minutes.

7. Ask students to share the strategies they used in worksheet steps 3 and 4. Students should be able to provide descriptions such as, *Any quadratic graph will have two x terms,* or, *I can tell I need a number when the line doesn't go through the origin.*

8. If students did not create the function $\frac{1}{x}$ in worksheet step 5, show them what this function would look like and ask them to explain why it doesn't look like the other functions they have created.

EXTEND

This activity is intended to provide students with a task in which they can practice both identifying and comparing graphs, and gain awareness of the symbolic makeup of different equations. It's a more concrete way of getting to the idea that an equation like $ax^2 + bx + c$ can generate any quadratic function—which would involve having a grab bag of $a, b, c, x, x, x, +, +, *, *$. Students might find it intriguing to see that those 10 ingredients can match any given quadratic. How many ingredients would be needed for all linear functions or all constant functions? What about all cubic functions?

ANSWERS

1. Possibilities include: $y = -1*x$, $y = x$, $y = x*x$, $y = x - 1$, $y = 4$, $y = x*x$

2. a. $y = 4$
   b. $y = -1*x$
   c. $y = x*x - 1$

3. a. The symbols $-$ and 1 were removed.
   b. The symbols $+$ and x were removed.
   c. The symbols $*$ and x were added.

4. Answers will vary, depending on how students interpret the scaling.

   a. There should be at least two x terms, as well as a multiplication sign and a positive number.

   b. There should be at least one x term, a subtraction sign, a multiplication sign, a negative number, and a positive number.

5. Answers will vary. Students might use the strategy of identifying whether the graph is linear or quadratic, and using that to decide how many x terms there will need to be. Some students might try to graph all the functions and deduce the needed items.

6. Answers will vary but should now include the exponential function $y = \frac{1}{x}$.

Grab Bag Graphs

In this activity you'll use a "grab bag" of symbols (numbers and operations) to generate as many different functions and graphs as possible.

EXPLORE

1. Suppose you are given a grab bag that contains the following items: 4, −1, +, *, *, x, x. You can use the items to create a function. Here are some valid functions: $y = 4 + x$, $y = x*x*4$, $y = -1 + 4*x$. Write down two more valid functions.

2. In a new sketch, choose **Graph | Plot New Function**. Using the items in your grab bag, enter expressions to produce the following graphs. You can double-click a function to change it.

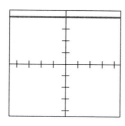 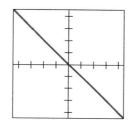 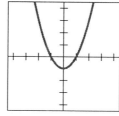

   a. $y =$ _____   b. $y =$ _____   c. $y =$ _____

3. In each of the following pairs of graphs, symbols were removed from or added to the first function in order to get the second one. Use Sketchpad to help you figure out what the symbols were and whether they were removed or added.

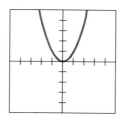 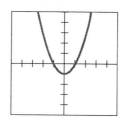

   a. _____

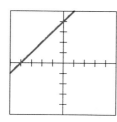 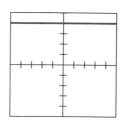

   b. _____

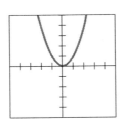 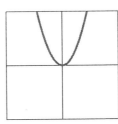 c. _____

4. Now make a different grab bag. Which items might be needed in the grab bag in order to create each of the following sets of graphs? Use Sketchpad to check.

a.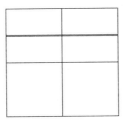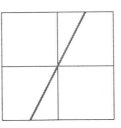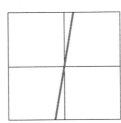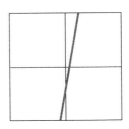

Grab bag includes _____

b.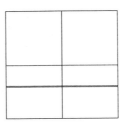

Grab bag includes _____

5. Describe the strategies you can use to figure out what items might be in a grab bag.

6. Suppose you have a grab bag containing *x, x, /, *,* and 1. Create as many functions as you can with these items. Sketch the graphs of your functions on another sheet of paper and label them.

EXPLORE MORE

7. Experiment with functions that have more than two *x* terms or that include parentheses. You may also want to investigate some of the functions that are found in the Functions pop-up menu of the Calculator, such as sine (sin), absolute value (abs), and logarithm (log).

Exploring Expressions and Equations in Grades 6–8 with The Geometer's Sketchpad
© 2012 Key Curriculum Press

Million Dollar Fortune:
Comparing Rates of Growth

ACTIVITY NOTES

INTRODUCE

Project the sketch for viewing by the class. Expect to spend about 15 minutes.

1. Open **Million Dollar Fortune.gsp** and go to page "Fortunes." Enlarge the document window so it fills most of the screen.

2. Explain, *Today you're going to compare different schemes for accumulating money and figure out which scheme would be best. You'll be able to use Sketchpad to help you figure out how much each scheme pays for a certain number of days. This will help you decide which scheme might be best under different circumstances. Before you begin, I'll demonstrate how the sketch works.*

3. Read the problem aloud. Ask students for suggestions of how they think different schemes could work. Try to get students to create schemes where money is paid each day. *Let's take a look at Fortune 1.* Press *Fortune 1* and read the fortune aloud. Ask the students to interpret the table that appears and to predict what the next row will look like. Then explain that they can select the table and press **+** on their keyboards in order to see the next row. Ask, *Estimate the total amount of money Fortune 1 will pay out after 10 days.* Write down a few of these predictions on the board.

4. Distribute the worksheet. Say, *Read the first step on your worksheet, which describes each of the four fortunes, and make a prediction about which fortune you think is best. Then answer the questions in steps 2 through 5.* This should take about 5 to 10 minutes. Circulate to make sure students have filled in their predictions and the different types of growth rates. If students have difficulty identifying the growth rate as quadratic (Fortune 3) or exponential (Fortune 1), they can simply describe which fortunes have the fastest and slowest growth rates.

DEVELOP

Expect students at computers to spend about 25 minutes.

5. Assign students to computers and tell them where to locate **Million Dollar Fortune.gsp.** Ask students to work from step 6 through step 12 and do the Explore More if they have time.

6. Let pairs work at their own pace. As you circulate, here are some things to notice.

   • Because only one table is shown at a time, help students develop strategies for comparing the different tables.

- Show students that they can also remove an entry from the table by selecting it and pressing the − key.

- Help students think of different situations for which one fortune might be better than another. For example, if for some reason money could be paid only on the first day, Fortune 4 is clearly the best. What are the constraints that determine when the other fortunes will be different?

SUMMARIZE

Project the sketch. Expect to spend about 5 minutes.

7. Bring students together to discuss their results. Ask students to compare their predictions to their answers in worksheet step 6. Students are often surprised to see how quickly the exponential growth increases. Draw attention to the fact that the best choice of fortunes depends on the number of days that money will be paid out.

8. If time permits, discuss the Explore More. Ask students to share their graphs for worksheet step 14. The graphs should show the general shapes of the curves. Students may find it challenging to label their axes properly because the values of the y-axis range from 0.01 to over 1 million. With the answers that students provide to worksheet step 13, you could choose plausible graphs for each fortune and graph each function on the same coordinate system to see whether it's possible to find the regions where the graphs intersect.

EXTEND

1. Challenge students to find the function expression that exactly models each fortune.

2. **What questions occur to you about types of growth rates?** Encourage curiosity. Here are some sample student queries.

If I were given a million dollars, I'd invest it and earn interest. Wouldn't it then be the best choice?

Is there a way to determine points of intersection exactly?

If Fortune 1 started with 2 cents, would it reach a million dollars in half the time?

Will exponential growth always outrun the other kinds of growth?

Are there kinds of growth rates other than these?

Is any of this kind of comparison useful outside of fortune cookies?

ANSWERS

1. Answers will vary.

2. Row 3: 0.04, 0.07. Row 4: 0.08, 0.15. Growth Rate: more than doubling (exponential)

3. Row 3: 50, 150. Row 4: 50, 200. Growth Rate: increasing by a constant (linear)

4. Row 3: 20, 45. Row 4: 25, 70. Growth Rate: accelerating (quadratic)

5. Answers will vary.

6. Most students will change the predictions. Fortune 4 is the best until day 26, when Fortune 1 exceeds 1 million dollars.

7. No. Fortune 2 is better than Fortunes 1 and 3 until day 16, but Fortune 4 is still the best at this point. Fortune 1 is better than Fortune 2 after day 15, and Fortune 3 and is better than Fortune 2 after day 16.

8. The graph increases very rapidly at an exponential rate.

9. Yes. The graph starts off with a greater y-intercept than that in Fortune 1, but does not increase as quickly. It increases at a constant rate to produce a linear graph.

10. Yes. The graph starts off with a y-intercept that is less than that in Fortune 2 and greater than that in Fortune 1. It increases faster than Fortune 2 and more slowly than Fortune 1 to produce a curved, quadratic graph.

11. Fortune 1 passes Fortunes 2 and 3 on day 16, and passes Fortune 4 on day 26. Fortune 3 passes Fortune 2 on day 17, and passes Fortune 4 on day 630. Fortune 2 passes Fortune 4 on day 19,999 (almost 50 years!).

12. In the long run, Fortune 1 pays out the most.

13. a. Corresponds to Fortune 2

   b. Corresponds to Fortune 4

   c. Corresponds to Fortune 3

   d. Corresponds to Fortune 1

14. Answers will vary. Look for graphs that have the right shape (Fortune 1 should be exponential, Fortune 2 should be linear, Fortune 3 should be quadratic, and Fortune 4 should be constant). All graphs should start at the y-axis and be in the first quadrant.

Exploring Expressions and Equations in Grades 6–8 with The Geometer's Sketchpad
© 2012 Key Curriculum Press

Million Dollar Fortune

 Name:

In this activity you'll use various techniques to try to identify the best of four fortune cookie schemes.

EXPLORE

1. After dinner at your favorite Chinese restaurant, you took four fortune cookies from the jar. Here are the fortunes from each cookie.

> **Fortune 1:** You'll be paid 1¢ today, and then each day after today you'll be paid twice the amount you were paid the day before.
>
> **Fortune 2:** You'll be paid $50 today, and then each day after today you'll be paid the same amount you were paid the day before.
>
> **Fortune 3:** You'll be paid $10 today, and then each day after today you'll be paid $5 more than the amount you were paid the day before.
>
> **Fortune 4:** You'll be paid $1 million today, and then that's it.

You must choose one fortune. Which should you choose? Write your prediction of which will pay you the most money.

2. For Fortune 1, compute the amount of money you'll receive on Days 3 and 4. Then use words to describe the growth rate of the total amount.

Day	Amount of Money Received (dollars)	Total Amount (dollars)	Growth Rate
1	0.01	0.01	
2	0.02	0.03	
3			
4			

3. For Fortune 2, compute the amount of money you'll receive on Days 3 and 4. Then use words to describe the growth rate or the total amount.

Day	Amount of Money Received (dollars)	Total Amount (dollars)	Growth Rate
1	50	50	
2	50	100	
3			
4			

4. For Fortune 3, compute the amount of money you'll receive on Days 3 and 4. Then use words to describe the growth rate of the total amount.

Day	Amount of Money Received (dollars)	Total Amount (dollars)	Growth Rate
1	10	10	
2	15	25	
3			
4			

5. For Fortune 4, you receive 1 million dollars. That's a lot of money! But how do you know it's the best choice? Based on your work so far, do you want to change your choice of a fortune?

 6. Open **Million Dollar Fortune.gsp** and go to page "Fortunes." Click each fortune cookie and read the fortune carefully. To increase the number of entries in the table, select the table and press the + key on your keyboard. Check your answers above. Then add rows to the tables as needed in order to help you decide which fortune to choose. Describe the reason for your choice of fortunes.

7. Is there ever a situation in which Fortune 2 is the best choice? Explain.

Exploring Expressions and Equations in Grades 6–8 with The Geometer's Sketchpad
© 2012 Key Curriculum Press

8. Go to page "Fortune 1." The graph contains the data points in the table, corresponding to the first few days of the fortune. Increase the number of entries in the table as you did in step 6 so that you can see the shape of the graph over time. You might want to change the vertical scale by dragging the numbers on the axis. Describe the general shape of the graph.

9. Go to page "Fortune 2." and repeat what you did in step 8. Compare the shapes of the two graphs you've produced. Will the two graphs ever intersect? Explain.

10. Go to page "Fortune 3." and repeat what you did in step 8. Compare all three graphs you've produced now. Will this graph ever intersect either of the other two? Explain.

11. Go to page "Compare." Press *Tables <--> Graphs* to switch between the two representations. Select all four tables and increase the number of entries in the tables at the same time. Use the graphs or tables to find the exact days when each fortune becomes a better choice than another fortune.

12. In the long run, which fortune is the best choice?

EXPLORE MORE

13. Describe which of the following equations would best model each of the graphs of the four functions. The letters *a, b,* and *c* represent numbers that would be constant for any particular equation.
    a. $y = ax + b$
    b. $y = a$
    c. $y = ax^2 + bx + c$
    d. $y = ab^x + c$

14. Try to create graphs for all of the four schemes on the same coordinate system so that you can see where they intersect. (*Warning:* It may be challenging to find an appropriate way to scale the axes.) Think about how you could change Fortune 2 or Fortune 3 so that it doesn't have the same value on Day 15. Find a way of changing all four fortunes so that they retain their basic shape, but all intersect at the same point.

Match It Up:
Comparing Graphs, Tables, and Equations

 ACTIVITY NOTES

INTRODUCE

Project the sketch for viewing by the class. Expect to spend about 10 minutes.

1. Open **Match It Up.gsp** and go to page "Graphs." Enlarge the document window so it fills most of the screen.

2. Explain, ***Today you're going to use Sketchpad to play a matching game in which you try to figure out the graphs or tables that correspond to a given equation. You'll be seeing linear, quadratic, and exponential equations of all kinds.*** Ask students to briefly describe each type of growth and give a few examples of each.

3. Ask, ***Which graph corresponds to the equation*** $y = x + 4$? [The third graph in the top row] Show students how they can check their own solutions by opening a new sketch and choosing **Graph | Plot New Function.** Tell students to press *New Equation* when they are ready for the next challenge. This button will randomly generate a new equation and a set of six different graphs. Tell them that later in the activity they will match an equation to tables. Warn them that sometimes none of the graphs or tables will match the equation.

4. If you want students to save the graphs that they plot, demonstrate choosing **File | Save As,** and let them know how to name and where to save their files.

DEVELOP

Expect students at computers to spend about 25 minutes.

5. Assign students to computers and tell them where to locate **Match It Up.gsp.** Distribute the worksheet. Tell students to work through step 5 and do the Explore More if they have time.

6. Let pairs work at their own pace. As you circulate, here are some things to notice.

   • Some equations do not have an associated graph or table. Encourage students to explain why none of the options they are given can be matched to the equation when that happens.

   • Encourage student to fill in as much detail as they can find in the third column of the worksheet tables. For example, if the equation is linear, can they identify the slope of the intercepts?

Exploring Expressions and Equations in Grades 6–8 with The Geometer's Sketchpad
© 2012 Key Curriculum Press

Match It Up: Comparing Graphs, Tables, and Equations
continued

 ACTIVITY NOTES

- Help students check their answers by graphing the given equation in a different sketch.

- You may want to encourage students to sketch a graph of the equation in worksheet step 3.

SUMMARIZE

Project the sketch. Expect to spend about 10 minutes.

7. Gather the class. Students should have their worksheets with them. Begin the discussion by opening **Match It Up.gsp** and use it to support the class discussion.

8. Start by asking students which equations were easiest to match. Students will probably find the constant, cubic, and exponential ones easier, as they have a characteristic form. Because there are different forms of quadratic equations, these equations may have been harder to match.

9. Go to page "Table." Generate a few new equations. Ask students to identify the correct table and also to describe what the corresponding graph would look like. If you have already generated one category of equation, invite students to tell you to skip to the next new equation until you have generated an example of each category.

10. Ask students whether they found the graphs or tables easier to match. Students might find that the graphs provide more immediate feedback, but that the tables provide specific values to check.

EXTEND

Have students use Sketchpad to plot functions. Explain, *So far the graphs have been presented on unnumbered axes. In a new sketch, use* **Graph | Plot New Function** *to re-create the graphs of any given equation. Then choose* **Graph | Grid Forms | Rectangular Grid** *so that you can change the scale of each axis independently by dragging the unit points or the numbers on the scales. This will allow you to explore how the graphs can change in appearance, depending on the scaling of the axes. Can you make it so that two graphs that started out looking different look more or less the same?*

ANSWERS

1. Answers will vary.

2. Answers will vary. Make sure students have one example of each type of equation and graph.

3. Answers will vary. Make sure students have one example of each type of equation and table.

4. Answers will vary.

5. Answers will vary. Accept solutions that are not quite exact, but that preserve the important properties of the graph.

Exploring Expressions and Equations in Grades 6–8 with The Geometer's Sketchpad

Match It Up

In this activity you'll match equations to their corresponding graphs and tables.

EXPLORE

1. Open **Match It Up.gsp** and go to page "Graphs." Press *New Equation.* Match the equation at the top of the page to one of the six graphs. Record the equation in the table, draw a sketch of the graph, and write a brief description of the specific attributes that helped you match them (the slope, the intercept(s), the general shape, and so on). Repeat at least four times.

Equation	Graph	Specific Attributes

2. Look back to the equations and graphs you found in step 1 and place them into categories according to whether they are linear, quadratic, cubic, or exponential. Show your categories in the space here. If you didn't find an example of each, press *New Equation* until you do, and write down the equation that corresponds to the missing category.

3. Go to page "Tables." Press *New Equation* and match the equation to one of the six tables. Keep track of the equations you match in the provided table. Copy one row of the table for each equation.

Equation	Example Row

4. Look back to the equations and tables you found in step 3 and place them into categories according to whether they are linear, quadratic, cubic, or exponential. Show your categories in the space here. If you didn't find an example of each, press *New Equation* until you do, and write down the equation that corresponds to the missing category.

5. Return to the "Graph" page. Press *New Equation.* Match your equation just as you did in step 1. Now, with the rest of the graphs, you'll also go in the reverse direction! Figure out what the equations are for the five other graphs on the sketch. Draw the graphs and equations in the table on the next page. When you are finished, compare your work with that of a partner.

Exploring Expressions and Equations in Grades 6–8 with The Geometer's Sketchpad
© 2012 Key Curriculum Press

Graph	Equation

EXPLORE MORE

6. In a new sketch, use **Graph | Plot New Function** to create some graphs of your own. Create three graphs using a variety of constant, linear, quadratic, cubic, and exponential functions. Hide the labels that show the equations of your three graphs. Challenge your partner to figure out which equation you used to create each graph.

Overlapping Squares:
Area and Perimeter Functions

INTRODUCE

Project the sketch for viewing by the class. Expect to spend about 5 minutes.

1. Open **Overlapping Squares.gsp** and go to page "Turn Squares." Enlarge the document window so it fills most of the screen.

2. Explain, *Today you're going to look at a simple geometric configuration and investigate how you can change the configuration to maximize its area and perimeter. As you can see, there are two congruent squares here on the sketch. The corner of the blue square has been pinned onto the middle of the yellow square, and you can rotate the blue square around that middle point by dragging point A.* Drag point *A* around several times. *You'll notice that the two squares overlap. Can anyone identify the shape of the overlap?* Students should be able to see that the overlap will change shape as you rotate the square. It's usually a quadrilateral. Sometimes it's a square and sometimes it's even a triangle.

3. *You're going to figure out how you can maximize the area of that overlapping part and how you can maximize the perimeter of the overlapping part. You'll be using both geometric and algebraic tools to help you answer these questions. Before you get started though, can anyone see any positions where the area or perimeter of the overlap will be the same?* The students should notice that as the blue square turns one full rotation, the overlap will look exactly the same four times.

DEVELOP

Expect students at computers to spend about 20 minutes.

4. Assign students to computers and tell them where to locate **Overlapping Squares.gsp.** Distribute the worksheet. Tell students to work through step 15 and do the Explore More if they have time.

5. Let students work at their own pace. As you circulate, here are some things to notice.

   • Worksheet steps 4–7 guide the students through the algebra needed to represent the area and perimeter in terms of a variable. Note that students are told to use *x*, which is defined to be half of *s*. This substitution avoids the repeated use of $\frac{1}{2}x$, which would produce unwieldy equations.

   • In worksheet step 6, students who aren't familiar with 45°-45°-90° triangles will need to apply the Pythagorean Theorem to find *BC*. Many students may need help to simplify the expression $\sqrt{2x^2}$. You might consider gathering the class at this point.

Exploring Expressions and Equations in Grades 6–8 with The Geometer's Sketchpad
© 2012 Key Curriculum Press

 ACTIVITY NOTES

- In worksheet step 7, you may need to help students reason about the area and perimeter of the overlap, using what they know about areas of triangles.

- In worksheet steps 8 and 9, students may be surprised that the area doesn't change. The *Show/Hide Triangles* button should help draw their attention to the two congruent triangles, one of which becomes smaller by the same amount that the other becomes larger, maintaining the area of the overlap.

- In worksheet steps 10–14, students use the angle of rotation on page "Measures" as the independent variable to explore the behavior of the perimeter and area. You may need to help students understand why it is sufficient to rotate the blue square only one-quarter of the way.

SUMMARIZE

Project the sketch. Expect to spend about 5 minutes.

6. Gather the class. Students should have their worksheets with them. Use **Overlapping Squares.gsp** to support the class discussion.

7. Invite students to explain why the area of the overlap does not change as point *A* moves, but why the perimeter of overlap does change. Students may have different ways of explaining it, so make sure to invite several explanations and not just one.

8. Ask students to describe what type of function the area and the perimeter of overlap have in terms of the angle. Challenge students to identify other relationships in the configuration. For example, the length of segment *BC* will change as the blue square rotates. The perimeter of the non-overlapping region in each square will also change.

EXTEND

1. It may be helpful to show students the graphs of the area and perimeter of the overlap as functions of the angle. On page "Measures," select the independent variable (angle), then the dependent variable (area or perimeter of the overlap), and choose **Graph | Plot as (x, y).** You will see a new coordinate system appear with the plotted point. As you drag point *A*, the plotted point will move accordingly. Select the plotted point and choose **Display | Trace Plotted Point** in order to see the shape of the graph. Alternately, select the plotted point and point *A*, and choose **Construct | Locus.** To make the shapes of the graphs easier to

see, choose **Graph | Grid Form | Rectangular Grid** and stretch the vertical axis.

2. ***What other questions might you ask about this configuration?***
Encourage all curiosity. Here are some ideas students might suggest.

How does the area of the overlap relate to the area of the squares?

What if the squares weren't congruent?

Would the same things happen if the squares were triangles?

ANSWERS

1. Predictions will vary.

2. *BC* and *CD* are both half as long as side *s*.

3. The relationships are true for any length of side *s*.

4. $BC = x$, $CD = x$

5. Area $= x^2$, perimeter $= 4x$

6. *BE* and *CE* are both equal to *x*. *BC* is about 1.4 times the length of *x*. Using the Pythagorean Theorem, $BC^2 = x^2 + x^2$, which simplifies to $BC = \sqrt{2}x \approx 1.41x$.

7. Area $= \frac{1}{2}(2x)(x) = x^2$, perimeter $= 2x + 2(\sqrt{2}x) \approx 4.82x$

8. The areas are equal.

9. The two shaded triangles are congruent because they have congruent bases and heights. As the blue square rotates, it loses an area equivalent to one shaded triangle, but at the same time, it gains an area equivalent to the other shaded triangle. This implies that the total area of the overlap remains the same as the square rotates.

10. The angle ranges from 0° to 90°. The area remains constant at 10.56 cm². The perimeter ranges from about 13.00 cm to 15.69 cm.

11. It doesn't matter how the blue square is positioned. The area does not change.

12. To maximize the perimeter, position the blue square so the angle measures either 0° or 90°. To minimize the perimeter, position the blue square so that the angle measures 45°.

Exploring Expressions and Equations in Grades 6–8 with The Geometer's Sketchpad
© 2012 Key Curriculum Press

13. The graph should be a horizontal line with a y-coordinate of 10.56 cm^2 and x-coordinates from 0° to 90°.

14. The graph should look like a parabola with a minimum at (45°, 13.00 cm). Again the x-coordinates should range from 0° to 90°. Students should be able to indicate the two maximum points at (0°, 15.69 cm) and (90°, 15.69 cm).

15. Answers will vary. Students should be able to argue that no matter what the size of the squares, the area of the overlap should never change, whereas the perimeter of the overlap will have its maximum and minimum at the same locations.

16. Both squares have side length 6.5 cm.

Overlapping Squares

 Name:

In this activity you'll investigate the area and perimeter of a changing geometric figure. Using both geometric and algebraic tools, you'll be able to figure out when the area and perimeter will achieve their maximum and minimum values.

EXPLORE

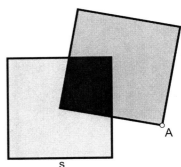

1. Open **Overlapping Squares.gsp** and go to page "Turn Square." Drag point A to turn the blue square around the center of the yellow one. Predict how you should position the squares to maximize the area and perimeter of the overlap.

2. Press *Show/Hide Lengths* and drag point A again. You should now see the length of side s, and in some positions you'll see other lengths as well. Press *Make AB parallel to side s.* How are BC and CD related to side length s?

3. Drag either endpoint of side s to change the size of the squares. If needed, press the button to make the bases of the squares parallel again. Are the relationships in step 2 still true? Experiment with other side lengths and describe what you notice.

4. To make later calculations easier, you'll now define a new variable, x, that is equal to half of side length s. Press *Define New Variable* and write expressions for BC and CD in terms of x.

5. How can you find the area and perimeter of the overlap when the bases of the squares are parallel? Use what you know about squares and your expressions from step 4 to express the area and perimeter in terms of x.

Exploring Expressions and Equations in Grades 6–8 with The Geometer's Sketchpad
© 2012 Key Curriculum Press

6. Press *Make AB pass through point D*. How are *BE* and *CE* related to *x*? How is *BC* related to *x*? Experiment with other side lengths. Then write an expression for *BC* in terms of *x*. (*Hint:* What do you know about △*BEC*?)

7. How can you find the area and perimeter of the overlap when one side of the blue square passes through a vertex of the yellow square? Use what you know about triangles and your expressions from step 6 to express the area and perimeter in terms of *x*.

8. Compare your expressions in steps 5 and 7. What do you notice about the area of the overlap for both positions?

9. Press *Show/Hide Triangles*. What do you notice about the two shaded triangles? Explain why the area of the overlap does not change as the blue square rotates.

10. Go to page "Measures." Drag point *A* and observe the values of the three measurements. Describe the range of each value.

11. How should you position the blue square to maximize the area, and to minimize the area?

12. How should you position the blue square to maximize the perimeter, and to minimize the perimeter?

13. Sketch a graph that describes the relationship between the angle and the area of the overlap.

14. Sketch a graph that describes the relationship between the angle and the perimeter of the overlap.

15. Was your prediction in step 1 correct? Would your answer change if the squares were bigger or smaller?

EXPLORE MORE

16. Use the measurements given on page "Measures" to figure out the size of the squares.

Exploring Expressions and Equations in Grades 6–8 with The Geometer's Sketchpad

Dancing Dynagraphs:
Exploring Quadratic Functions

ACTIVITY NOTES

INTRODUCE

Project the sketch for viewing by the class. Expect to spend about 5 minutes.

1. Open **Dancing Dynagraphs.gsp** and go to page "Dynagraphs." Enlarge the document window so it fills most of the screen.

2. Explain, *Today you're going to use Sketchpad to explore functions in a new way. Instead of using tables or coordinate graphs, you'll work with dynagraphs. In your work with linear functions, you have studied the slope and have seen how the slope determines the steepness of the graph. The slope also describes the constant rate at which the dependent variable y changes with respect to the independent variable x. This way of describing slope focuses on the behavior of linear functions. In this activity you'll investigate the behavior of quadratic functions. You'll try to figure out how the dynagraph gives you information about the special characteristics of quadratic functions and you'll compare the behavior of linear and quadratic functions. Before you begin, I'll demonstrate how the dynagraph works.*

3. If necessary, press *Dynagraph 1*. Illustrate how you can drag point x on the top axis and that doing so will change the location of point y. As you drag x, ask, **Where should I drag x so that y is equal to −5?** The students should be able to notice that no matter where you drag x, y never seems to drop below 0.

4. *As you work on this activity, you'll also be asked to explore four other dynagraphs.* Show students how to click the other buttons on the sketch in order to change dynagraphs.

DEVELOP

Expect students at computers to spend about 20 minutes.

5. Assign students to computers and tell them where to locate **Dancing Dynagraphs.gsp.** Distribute the worksheet. Tell students to work through step 3. As you circulate, here are some things to notice.

- Students may have trouble knowing what is meant by "special" values in worksheet step 1. Encourage them to look for values of x that cause an important change in the behavior of y. Some students will become aware of the values of x that produce a minimum or maximum for y. Other students might also pay attention to the values of x that make y go off the screen, but these are not as mathematically interesting. You

might help students see that changing the size of the window would change the latter values, whereas the former values are independent of the size of the window (or the scale of the axes).

- Encourage students to describe more qualitatively the way in which *y* moves as they drag *x*. They should be able to see that as *x* moves away from its "special" point, *y* seems to speed away, or "shoot off." Students should at least notice that *y* does not stay a constant distance away from *x*, nor does the distance seem to simply double or triple (grow at a constant rate) as *x* moves.

6. Gather students together to compare their solutions to worksheet steps 1–3. Refer to the solutions to help guide your discussion. Students should be able to describe some similarities between the five dynagraphs. ***Now we'll look at a whole range of quadratic dynagraphs and compare them with some linear functions too.*** Assign students the remainder of the steps.

7. The goal of steps 4–6 is for students to be able to link the algebraic form they have seen in quadratic equations to the differences in behavior between linear and quadratic dynagraphs. As you circulate, make sure students are able to describe the difference between linear and quadratic dynagraphs in the way *y* changes as they drag *x*. Also encourage students to experiment with a wide range of different values for *a*.

SUMMARIZE

Project the sketch. Expect to spend about 5 minutes.

8. Gather the class. Students should have their worksheets with them. Use **Dancing Dynagraphs.gsp** to support the class discussion.

9. Go to page "Equations." Begin with worksheet step 6. Ask students to describe the effect that *a* has on the behavior of the dynagraph. Make sure to illustrate examples that students propose by entering their value for *a* and dragging *x*.

10. Ask students to describe the difference between the behavior of the dynagraph when $a = 0$ and when $a \neq 0$. The students should be able to see that *y* does not grow at a constant rate when $a \neq 0$. It may be difficult to see slower rates of growth, but students should be able to see faster ones. Relate this behavior to both the algebraic equation (the x^2 term will have a varying effect on the equation) and to the graphical representation (the quadratic function produces curved graphs rather

than straight ones—sometimes these curves are shallow and sometimes they can be quite steep). Tell students, *Although we often work with the algebraic or graphical forms of the quadratic functions, it's often useful when you are trying to model a phenomenon to first identify the basic behaviors involved. As you continue to study different functions, you will encounter the exponential function, which grows at a proportional rate and which seems to fit real-world phenomena better than the quadratic function does.*

EXTEND

1. Tell students, *Add a new page to the document you have been working with and explore the graphical counterparts to some of the quadratic functions on the "Equation" page. Choose* **Plot New Function** *from the Graph menu and enter one of the equations you created on the "Equation" page (for example, you might try $x^2 + 3x + 1$). Compare the graphs you obtain with the way the corresponding dynagraphs behave. Which representation of the function do you find more interesting or useful?* Help students create their function plots. Encourage them to create several of them (and to label the different ones). Students might find the dynagraphs more interesting because they move and change, but the graphs they create may provide more useful information (the vertex, the roots, the general shape).

2. What questions occur to you about quadratic functions and rates of change? Encourage curiosity. Here are some sample student queries.

   Why do the changing y-values of quadratics reverse direction?

   Is there an easy way to see from a quadratic equation where its maximum or minimum is?

   How would dynagraphs of cubic functions behave?

ANSWERS

1. Answers will vary. The important special value of x is the one that causes y to change direction (which corresponds to the minimum or maximum value of y).

2. Similarities: They all have a restricted range, but an unrestricted domain; they all seem to change direction at a special value of x; they

all have a special value; and y seems to speed away as x moves away from the special point. Differences: Sometimes the $f(x)$ seems to have a minimum and sometimes a maximum.

3. Answers will vary. The value of y does not change at a constant rate as x is dragged. This can be seen by comparing the change in y over any two equal intervals in the change of x. This is particularly noticeable when the rate of change is positive in one interval and negative in the other.

4. In this dynagraph the range is no longer restricted and there are no special values. Also, y grows at a constant rate (it's always about 3 times as large as x).

5. When $a = 0.00$, the function becomes linear (there is no longer an x^2 term). So the dynagraph behaves as it would with any other linear function.

6. Answers will vary. When $a \neq 0$, the function is quadratic. When $a > 0$, the dynagraph will show a minimum value for y, and when $a < 0$, the dynagraph will show a maximum value for y. The greater the absolute value of a, the faster y seems to speed away as you drag x.

7. Answers will vary. For example, setting x to 0 will allow students to find the value of c.

Dancing Dynagraphs

 Name:

In this activity you'll investigate a new way of seeing and interpreting quadratic functions.

EXPLORE

1. Open **Dancing Dynagraphs.gsp** and go to page "Dynagraphs." Drag point x across the top axis. What happens to point y? Identify special values of x where something interesting happens to y.

2. Press *Dynagraph 2.* Drag x again. Do the same for each of the five dynagraphs. What do the five dynagraphs have in common? How are they different?

3. Using any one of the five dynagraphs, explain whether y seems to change at a constant rate. It may help to look carefully at the values of x and y as you drag.

4. Go to page "Equation." Drag x along the axis. Describe how this dynagraph differs from the ones on page "Dynagraphs".

5. Press *Show Controls.* You'll be able to see the equation for the relationship between x and y. Explain how the value of $a = 0.00$ in the equation relates to the behavior you noticed above.

6. Change the value of a by double-clicking it and entering a new value. Describe how the value of a changes the way y behaves as x moves along the axis.

Dancing Dynagraphs

continued

EXPLORE MORE

7. Choose values for the coefficients *a, b,* and *c.* Then press *Hide Controls.* Challenge a classmate to tell you as much as possible about your equation based on the dynagraph. Think about whether it's possible to discern a difference in behavior for each of the coefficients.

Exploring Expressions and Equations in Grades 6–8 with The Geometer's Sketchpad
© 2012 Key Curriculum Press

Equations and Inequalities

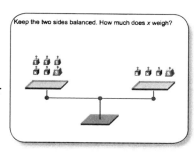

Balancing: Solving Linear Equations

Students use a Sketchpad pan balance model to solve a sequence of equations of increasing difficulty. They transfer the manipulation of the pan balance to solving equations independent of the balance.

Balancing with Balloons: Solving Equations with Negatives

Students use a Sketchpad pan balance model to solve a sequence of equations with positive numbers and variables, represented by weights, and negative numbers and variables, represented by balloons. As the problems increase in difficulty, students transfer the manipulation of the pan balance to solving equations independent of the balance.

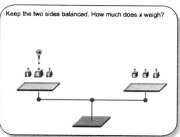

Undoing: Solving Linear Equations

Students develop strategies for undoing a sequence of operations to solve simple linear equations. They drag operations and change values in a model so that a two-step process on the bottom undoes a two-step process on the top.

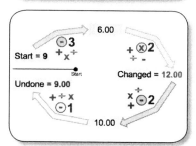

Levers and Loads: Solving Linear Inequalities

Students use a model of a lever to investigate a problem involving the force required to lift an object. Using an equation that relates forces and their distances from the fulcrum, they learn to derive and solve a linear inequality.

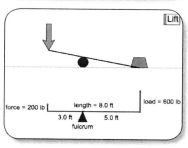

Shady Solutions: Graphing Inequalities on a Number Line

Students use a prepared sketch to graph the solutions to inequalities on a number line. In addition to the relationship between the solution graphs and the algebraic techniques for solving inequalities, students examine why solving an inequality involving the opposite of x results in reversing the inequality symbol.

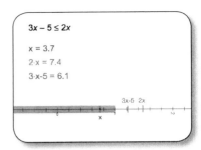

In or Out: Graphing Inequalities

Students drag a special point on a coordinate grid to create the region that represents the solution to a linear inequality. They investigate the relationships among inequalities, regions on the coordinate grid, solution points, and the boundary line.

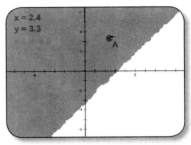

Around We Go: Equation of a Circle

Students observe the relationship between the coordinates of points on a circle. By applying the Pythagorean Theorem, the activity leads students to develop the equation of a circle.

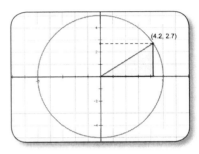

Make It Move: Distance-Rate-Time Animations

Students deepen their understanding of distance-rate-time problems by creating animated models. After making scale drawings by hand to model and solve problems, students animate their drawings in dynamic sketches, test their solutions, confirm their approaches, consider any mistakes that are revealed, and revise and retest as needed.

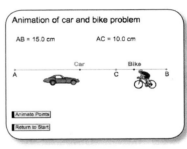

Exploring Expressions and Equations in Grades 6–8 with The Geometer's Sketchpad
© 2012 Key Curriculum Press

Balancing: Solving Linear Equations

 **ACTIVITY NOTES**

INTRODUCE

Project the sketch for viewing by the class. Expect to spend about 10 minutes.

1. Open **Balancing.gsp** and open page "Balance." Enlarge the document window so it fills most of the screen.

2. Explain, *Today you're going to use a Sketchpad balance to model solving equations. Have you ever seen a balance like this one? How does it work?* Let students share their understanding of balances. Some may have had experience in science classes using a triple-beam balance, which has a pan on one side and calibrated weights on the other, but few have probably used a balance with two pans. You might compare it to a seesaw, which will be familiar to most students. Make sure students understand that the heavier side will go down and that the two sides will balance only if they have the same weight.

3. Demonstrate how the model works, using only numbered weights. Remove a weight from the one side of the balance by dragging it into the storage bin. Then add some weights by dragging them from the storage bin. *What do you see?* [The balance tilts down toward the heavier side.]

4. Drag one x-weight to the left side and three 1-weights to the right side. They should balance (x is set equal to 3 on this page). *What does this tell you about the weight of x?* [It equals 3.] *How would you write this as an equation?* [$x = 3$] Use the **Text** tool to write the equation so that the equal sign is aligned with the center of the balance. Use a large font for visibility.

For a new text caption, use the **Text** tool. To move a caption, use the **Arrow** tool. To edit a caption, either click once with the **Text** tool or double-click with the **Arrow** tool.

5. Change back to the **Arrow** tool and add one 5-weight to the left side. *What can I do to the right side to keep the balance?* After students suggest adding 5, drag a 5-weight to the right side and write the new equation, $x + 5 = 8$. Use a new text caption so that you can align the equal signs.

6. Select the original equation and choose **Display | Hide Caption.** Then ask, *What if we were trying to solve this equation? What could we do to isolate x, that is, to get x by itself?* [Subtract 5 from both sides.] You want students to see that the process of solving an equation is the opposite of building one. Remove both 5-weights and unhide the original equation.

7. Add another x-weight on the left and ask, *What can I do to the right side to keep the balance?* Some students may suggest adding an x to the other side (adding the same thing to both sides) and others may suggest

adding three 1's to the other side (doubling both sides). Model both suggestions in turn, changing the caption to show the new equation ($2x = x + 3$, or $2x = 6$), and clarifying these two different interpretations. Show that to solve the equation $2x = 6$, students must reverse the process of doubling by finding half of what's on each side. ***What mathematical operation does this model?*** [Division by 2]

8. Go to page "A." ***Today you'll be solving equations like these. What is the original equation?*** [$x + 3 = 5$] Use the **Text** tool to write the equation in the sketch, and emphasize that students will do this on their worksheets. Then say, ***Your goal is to figure out how much x weighs. But you must do this while keeping the two sides balanced. What can you do to both sides of the balance to make it simpler?*** [Subtract three 1's from both sides.] Make sure students understand the idea of performing the same operation on both sides of the equation.

9. Say, ***The problems get more complex as you go, and you'll have to do operations besides subtraction. Some will require more than one step. Be sure to record the equations and the solution steps on your worksheets.*** If you wish to give more guidance, the page "Example" has an animation of how to solve the equation $3x + 2 = 8$ (press *Show Steps*). You might do the example together with the class, ask student pairs to review it on their own, or mention it to student pairs only if they need support.

DEVELOP

Expect to spend about 25 minutes.

10. Assign pairs to computers and tell them where to locate the sketch **Balancing.gsp.** Distribute the worksheet. Tell students to work through step 7 and do the Explore More if they have time.

11. Let pairs work at their own pace. As you circulate, here are some things to notice.

   • In worksheet step 3, equation B is designed to give students practice replacing a 5-weight with five 1-weights. Otherwise it is exactly like equation A.

   • In worksheet steps 4 and 5, check that students understand that equation C requires division whereas equations A and B involved subtraction.

Exploring Expressions and Equations in Grades 6–8 with The Geometer's Sketchpad
© 2012 Key Curriculum Press

 ACTIVITY NOTES

- In worksheet steps 6 and 7, watch that students are systematically doing the same thing to both sides of the balance. Starting with equation D, the first two-step equation, some students might go directly to moving all but one x from one side and then manipulating the other side until they find the solution. This will work, but misses the point. Encourage students to articulate what they're doing in terms of mathematical operations (subtracting and dividing both sides of the equation by the same thing).

- Check that students are recording the equations and their solution steps on their worksheets.

- Some students are likely to work through the equations more quickly than others. Encourage them to experiment with the Explore More to build equations of their own.

SUMMARIZE

Expect to spend about 10 minutes.

12. Bring the class back together to discuss the strategies they discovered. ***What have you learned about solving equations?*** Bring out these objectives.

   - Equations might be solved by balancing, similar to working with a pan balance.

   - Subtracting the same number from both sides of an equation does not change the solution.

   - Dividing both sides of an equation by the same number does not change the solution. (You might remind students that they cannot divide by zero.)

13. ***What other questions can you ask that you may or may not be able to answer?*** Encourage all student curiosity. Mathematical questions of interest include these.

   - Can you also add and multiply both sides of an equation by the same thing?

   - Can you use balancing if the numbers aren't whole numbers?

   - Why do we need this balancing method if we can just guess and check solutions?

   - For what kinds of equations can we not guess and check solutions?

EXTEND

For practice with solving simple linear equations without the balance, randomly generated linear equations appear on page "Practice." Students who already have strong equation-solving skills should enjoy creating complex equations of their own on the page "Balance."

ANSWERS

1. $x + 3 = 5$

2. $x = 2$

3. $x + 3 = 7$
   $x = 4$

4. Even if you replace the 5-weight with five 1-weights, you must use division to find the solution.
   $5 = 2x$
   $2.5 = x$

5. Equation C modeled division. Equations A and B modeled subtraction.

6. Equation D: $2x + 1 = 5$
   $2x = 4$
   $x = 2$

7. Equation E: $2x + 8 = 3x + 5$
   $8 = x + 5$
   $x = 3$

   Equation F: $4x + 1 = x + 7$
   $3x + 1 = 7$
   $3x = 6$
   $x = 2$

   Equation G: $x + 7 = 4x + 4$
   $7 = 3x + 4$
   $3 = 3x$
   $x = 1$

8. Answers will vary.

9. Answers will vary.

   Exploring Expressions and Equations in Grades 6–8 with The Geometer's Sketchpad
© 2012 Key Curriculum Press

Balancing

 Name:

In this activity you'll use Sketchpad's balance model to solve equations.

EXPLORE

1. Open **Balancing.gsp.** Go to page "A." Write the equation represented by the balance.

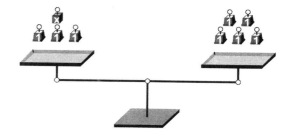

2. To isolate the *x*, move 3 units off of each side of the balance. Does the scale still balance? Write the resulting solution.

3. Go to page "B." Write the equation. Drag items while keeping the two sides balanced to find the solution. Write the solution.

   Original equation:

   Solution:

4. Go to page "C." Write the equation. Explain why you can't drag items to find the solution. How can you find the solution for *x*?

5. What you did in equation C modeled a different operation than what you did in equations A and B. Which operations did you use to solve C? Which did you use to solve A and B?

6. Go to page "D." Solve this equation in two steps. After each step, the scale should balance. Record the balanced equations at each step here.

   Original equation:

   After first step:

   Solution:

7. Equations E, F, and G each require two or more steps. Use the balance to solve them and record your steps.

   Equation E:

   Equation F:

   Equation G:

EXPLORE MORE

8. Go to page "Balance." Build your own equation by putting *x*-weights and numbered weights on each side of the balance. Add 1-weights as necessary to balance the scale. Now solve your own equation, or challenge a classmate to solve it. When you're done, press *Change x-value* and repeat.

9. In the sketch, the *x*-value is always a whole number. Write an equation whose solution is a fraction. Show the solution steps.

10. Go to page "Practice" and try solving equations without using the model. Check you work by pressing *Show Solution;* then press *New Equation* for more practice.

Exploring Expressions and Equations in Grades 6–8 with The Geometer's Sketchpad
© 2012 Key Curriculum Press

Balancing with Balloons:
Solving Equations with Negatives

INTRODUCE

Project the sketch for viewing by the class. Expect to spend about 10 minutes.

1. Students should have already completed the activity Balancing, which introduces the Sketchpad balance model using only positive weights. If not, ask, **Have you ever seen a balance like this one? How does it work?** Make sure students understand that the heavier side will go down and that the two sides will balance only if they have the same weight. You should also show examples involving only positive weights before introducing the balloons.

2. Open **Balancing with Balloons.gsp** and go to page "Balance." Enlarge the document window so it fills most of the screen. Explain, **Today you're going to use a Sketchpad balance to model solving equations that include negative numbers and negative variables. Now there are balloons that pull up on the balance in addition to the weights that push down on the balance.** Drag a 1-weight to one side of the balance, and then drag a −1-balloon to the same side. **A negative one balloon pulls up just as much as a positive one-weight pushes down, so the scale remains balanced.** Demonstrate that the same is true for an x-weight and a −x-balloon.

3. Drag one x-weight to the left side and four 1-weights to the right side. They should balance (x is set equal to 4 on this page). **What does this tell you about the weight of x?** [It equals 4.] **How would you write this as an equation?** [$x = 4$] Use the **Text** tool to write the equation so that the equal sign is aligned with the center of the balance. Use a large font for visibility.

For a new text caption, use the **Text** tool. To move a caption, use the **Arrow** tool. To edit a caption, either click once with the **Text** tool or double-click with the **Arrow** tool.

4. Add a −1-balloon on the left side and ask, **What can I do to the right side to keep the balance?** Students may suggest adding a −1-balloon to the right or removing a 1-weight from the right. Acknowledge both approaches, and then say, **What happens if I do add a negative one balloon to the right?** Besides balancing the scale, elicit the idea that a 1-weight and a −1-balloon cancel each other out. Drag the balloon so it is directly above the weight, select both of their points, and drag them simultaneously out to the storage bin.

5. Write the equation $x − 1 = 3$. Use a new text caption so that you can align the equal signs. Hide the original equation and ask, **What if we were trying to solve this equation? What could we do to isolate x, that is, to get x by itself?** [Add 1 to both sides.] Drag a 1-weight to both sides

of the equation. Then select both the weight and balloon on the left side, remove them, and unhide the original equation.

6. ***Today you'll be solving equations like these. What is the original equation? The problems get more complex as you go, and most will require more than one step. Be sure to record the equations and the solutions steps on your worksheets.*** If you wish to give more guidance, the page "Example" has an animation of how to solve the equation $-x + 4 = 2x - 2$ (press *Show Steps*). You might do the example together with the class, ask student pairs to review it on their own, or mention it only to student pairs if they need support.

DEVELOP

Expect to spend about 25 minutes.

7. Assign pairs to computers and tell them where to locate the sketch **Balancing with Balloons.gsp.** Distribute the worksheet. Tell students to work through step 6 and do the Explore More if they have time.

8. Let pairs work at their own pace. As you circulate, here are some things to notice.

   • Watch that students are systematically doing the same thing to both sides of the balance. Some students might go directly to moving all but one x from one side and then manipulating the other side until they find the solution. This will work, but misses the point. Encourage students to articulate what they're doing in terms of mathematical operations (subtracting and dividing both sides of the equation by the same thing).

   • In worksheet step 4, equation C requires students to replace a -5-balloon with five -1-balloons in order to divide it into two groups. Students may also just do the division mentally, which is fine.

   • In worksheet step 5, check that students understand that the solution to equation C is negative. The problem asks what this means in terms of the model. One way of thinking about it is that the x's are balloons, which would mean that $-x$-balloons are weights. The main point of the question is to make students think, but don't focus too much on this limitation of the model.

   • In worksheet step 6, students will need to remove $-x$-balloons. Generally, students will find it easier to use positive x-weights to cancel out $-x$-balloons. If students don't, they will need to divide

Exploring Expressions and Equations in Grades 6–8 with The Geometer's Sketchpad
© 2012 Key Curriculum Press

 ACTIVITY NOTES

by a negative number on the last step, which is difficult to explain with this model. One way to think about it is that you isolate one $-x$-balloon, and so the solution for x must be the opposite of that value.

- Some students are likely to work through the equations more quickly than others. Encourage them to experiment with the Explore More to build equations of their own.

SUMMARIZE

Expect to spend about 10 minutes.

9. Bring the class back together to discuss the strategies they discovered. ***What have you learned about solving equations?*** Bring out these objectives.

- Equations might be solved by balancing, similar to working with a pan balance.

- Adding and subtracting the same number from both sides of an equation does not change the solution.

- Dividing both sides of an equation by the same number does not change the solution. (You might remind students that they cannot divide by zero.)

10. ***What other questions can you ask that you may or may not be able to answer?*** Encourage all student curiosity. Mathematical questions of interest include these.

- What does it mean to have a negative value for the solution?

- Can you also multiply both sides of an equation by the same thing?

- Can you use balancing if the numbers aren't integers?

EXTEND

For practice with solving simple linear equations without the balance, randomly generated linear equations appear on page "Practice." Students who already have strong equation-solving skills should enjoy creating complex equations of their own on page "Balance."

ANSWERS

1. $x - 5 = 2$

2. Add a 5-weight to both sides, and then remove the 5-weight and -5-balloon from the left side.

   $x = 7$

3. $3 = 2x - 1$

   $4 = 2x$

   $x = 2$

4. $2x + 1 = -5$

   $2x = -6$

   $x = -3$

5. The solution to equation C is negative. In terms of the model, x's are balloons (and $-x$-balloons are weights). Mostly, this illustrates a limitation of the model.

6. Equation D: $7 - x = 2x + 1$

   $7 = 3x + 1$

   $6 = 3x$

   $x = 2$

   Equation E: $x - 1 = -3x - 5$

   $4x - 1 = -5$

   $4x = -4$

   $x = -1$

   Equation F: $-3x + 2 = -x + 10$

   $2 = 2x + 10$

   $-8 = 2x$

   $x = -4$

   Equation G: $-x - 11 = -4x - 2$

   $3x - 11 = -2$

   $3x = 9$

   $x = 3$

7. Answers will vary.

8. Answers will vary.

Balancing with Balloons

 Name:

In this activity you'll use Sketchpad's balance model to solve equations that include negative numbers and negative variables, represented by balloons.

EXPLORE

1. Open **Balancing with Balloons.gsp** and go to page "A." Write the equation represented by the balance.

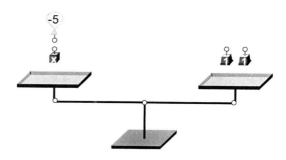

2. How can you isolate the *x*? Explain and write the resulting solution.

3. Go to page "B." Solve this equation in two steps. After each step, the scale should balance. Record the balanced equations at each step.

   Original equation:

   After first step:

   Solution:

4. Go to page "C." Solve this equation in two steps. After each step, the scale should balance. Record the balanced equations at each step.

   Original equation:

   After first step:

   Solution:

5. How is the solution to equation C different from the solutions to equations A and B? What does this mean in terms of the model?

6. Equations D, E, F, and G each require two or more steps. Use the balance to solve them and record your steps.

   Equation D:

   Equation E:

   Equation F:

   Equation G:

EXPLORE MORE

7. Go to page "Balance." Build your own equation by putting weights and balloons on each side of the balance. Add 1-weights and –1-balloons as necessary to balance the scale. Now solve your own equation, or challenge a classmate to solve it. When you're done, press *Change x-value* and repeat.

8. In the sketch the *x*-value is always a whole number. Write an equation whose solution is a fraction. Show the solution steps.

9. Go to page "Practice" and try solving equations without using the model. Check your work by pressing *Show Solution,* then press *New Equation* for more practice.

Undoing: Solving Linear Equations

 ACTIVITY NOTES

INTRODUCE

Project the sketch for viewing by the class. Expect to spend about 10 minutes.

1. Open **Undoing.gsp** and go to page "Explore." *What calculations are being made in this diagram?* [On the top branch, 3 is subtracted from 9, and the resulting 6 is multiplied by 2 to get 12. On the bottom branch, the 2 is subtracted from 12, and then 1 is subtracted from the result to get back to 9.]

2. Show how to drag one of the operations off the top branch and replace it with a different operation. *What happened?* [The bottom branch no longer undoes the top branch to get back to 9.]

3. Replace the original operation to return the diagram to the original form. Then demonstrate how to double-click on one of the numbers in the top branch and change its value. Again, the calculations on the bottom branch no longer result in the starting number.

DEVELOP

Expect students at computers to spend about 25 minutes.

4. Assign student pairs to computers and tell them where to locate **Undoing.gsp.** Distribute the worksheet. Ask students to work through step 8 and to go on to Explore More if they have time. Encourage students to ask each other for help with Sketchpad.

5. Let pairs work at their own pace. As you circulate, here are some things to notice.

   - In worksheet step 1, students should describe their reasoning, not simply state, "It doesn't work for all numbers."

   - If students pull the slider out very far to change the start number, they may have trouble getting it back onto the screen. Suggest they use **Edit | Undo.**

   - If students happen to choose **Display | Show All Hidden,** they should hide everything before they do anything else. If they have already clicked somewhere, they should undo back to the point where the controls were hidden.

   - In worksheet step 2, you might recommend that students begin with *Changed* and end with *Start.*

- You might suggest for worksheet step 3 that some students set up the top branch and challenge other students to find the corresponding bottom branch.

6. The Explore More activity extends undoing to equations whose coefficients or solutions are negative or fractions or decimals.

SUMMARIZE

Expect to spend about
10 minutes.

7. Reconvene the class. Students should have their worksheets with them.

8. *What strategies did you find?* Students likely will have found a successful strategy, but they may have difficulty articulating it. Use the terms *inverse operation* and *opposite order*. Reach class consensus on the main objective of the lesson: ***Some equations can be solved by undoing—considering the operations in opposite order and replacing each operation with its inverse operation.***

9. *Does undoing equations remind you of anything in your everyday life?* [Sample answer: Putting on and taking off shoes and socks—you put on socks before putting on shoes; to undo, you remove shoes before removing socks.]

10. *What other questions might we ask?* Encourage all student curiosity. Mathematical questions of interest students might raise include these.

   When page "Explore" was in its original format, why did Undone *keep changing by 2 when* Start *was changing by 1?*

   What's undoing good for?

   Are there equations that can't be solved this way?

Exploring Expressions and Equations in Grades 6–8 with The Geometer's Sketchpad
© 2012 Key Curriculum Press

ANSWERS

1. Answers will vary. The bottom branch does not undo the top branch for any starting number except 9. Each time you change *Start* by 1, *Undone* changes by 2.

2. $(Changed \div 2) + 3 = Start$

3. Answers will vary. The bottom equation should undo the top one.

4. Answers may vary. The operations will be considered in reverse order and replaced by their inverse operations.

5. Bottom: $(31 + 9) \div 5 = Start$, or $(31 + 9) \div 5 = 8$

    $Start = 8$

    Success of strategy and revised strategy may vary.

6–7. Answers will vary. The bottom equation should undo the top one, and the value of *Start* should check in the top equation.

8. a. $x = 1$

    b. $x = 6$

    c. $y = 17$

9. a. $x \approx 4.43$

    b. $x = -5$

    c. $n = \dfrac{13}{3}$

    d. $y = -8.48$

Undoing

In this activity you'll undo a linear equation to find an unknown value. You'll develop the strategy with known values before using it to find unknowns.

EXPLORE

1. Open **Undoing.gsp** and go to page "Explore." The calculations on the bottom branch of the diagram transform the 12 back into the start number, 9.

   The goal is not just to get back to 9. You want the bottom branch to undo the top branch for any start number. Drag the value of *Start*. Does the bottom branch undo the top branch for all start numbers? What patterns do you see?

2. On the bottom branch, drag the old operations out, drag new operations in, and change the numbers by double-clicking them until you get *Undone* to always match *Start* even when you drag the slider.

   The top branch can be represented by the equation

   $$(Start - 3) \times 2 = Changed$$

   What equation can represent your bottom branch?

3. Change the pink and blue operations and numbers on the top branch to set up a new situation. Then change the bottom branch until the undone number matches the start number for all start numbers.

   Top equation:

   Bottom equation:

4. What strategy do you think will always make the bottom branch undo the top?

Exploring Expressions and Equations in Grades 6–8 with The Geometer's Sketchpad
© 2012 Key Curriculum Press

5. To test your strategy, go to page "Solve." The start value is no longer visible. The equation for the top branch is (*Start* × 5) − 9 = 31. Follow your strategy to drag in operations and change the numbers in the bottom branch to undo the top branch. Then press *Show Start* to see whether your result equals *Start*. Record your results.

   Bottom equation:

   Value of *Start*:

   Was your strategy successful? If not, what is your revised strategy?

6. Press *New Problem* and repeat step 5. Record your results. Use the letter *x* in place of the word *Start*.

   Top equation:

   Bottom equation:

   Value of *x*:

7. Press *New Problem* and repeat step 5. Record your results. Use the letter *x* in place of the word *Start*.

   Top equation:

   Bottom equation:

   Value of *x*:

8. Solve these equations. Use page "Explore" of the sketch if needed, but feel free to solve them without using the sketch. Write the equation that undoes the first one and calculate the value of the variable that represents the *Start* value.

   a. $(x + 6) \times 5 = 35$ b. $(x \div 3) + 7 = 9$ c. $(y - 5) \div 6 = 2$

EXPLORE MORE

9. The undoing process can solve equations even if the numbers are not positive integers. Solve these equations. Write each undoing equation and calculate the starting value, the value of the variable.

    a. $(x \times 3.5) + 2.1 = 17.6$ b. $(x + 7) \div 2 = 1$

    c. $\left(n - \dfrac{1}{3}\right) + \dfrac{7}{3} = \dfrac{19}{3}$ d. $\dfrac{y - 4.3}{7.1} = -1.8$

Exploring Expressions and Equations in Grades 6–8 with The Geometer's Sketchpad
© 2012 Key Curriculum Press

Levers and Loads:
Solving Linear Inequalities

ACTIVITY NOTES

INTRODUCE

Project the sketch on a large-screen display. Expect to spend about 5 minutes.

1. Open **Levers and Loads.gsp** and go to page "Problem." Ask a volunteer to read the problem aloud. To help students make sense of the problem, you might ask them to draw a diagram.

2. Take a vote of "yes," they can lift the rock with a lever, or "no," they cannot. Students might show thumbs up or down. Record the totals on the board.

3. You or a student can demonstrate how to change the weights and how to move the fulcrum. Be careful not to move the fulcrum within 2 feet of the load. Press *Lift*. Note that the rock wasn't really lifted.

DEVELOP

Expect students at computers to spend about 20 minutes.

4. Assign students to computers and tell them where to locate **Levers and Loads.gsp.** Distribute the worksheet. Tell students to work through step 13 and do the Explore More if they have time. Encourage students to ask their neighbors for help if they are having difficulty with Sketchpad.

5. Students may have difficulties with worksheet steps 5 or 6. If so, convene the class and have students share ideas. Then let them continue with worksheet steps 7–13.

6. Let pairs work at their own pace. As you circulate, here are some things to notice.

   • You may need to remind students how to distribute. *One of the factors you are multiplying has two terms. What will you do?*

   • You can review solving linear equations with students who need help. *You can write equivalent equations to keep the sides balanced by doing the same thing to both sides. What operation could you use here?*

   • In worksheet step 12, encourage creative (but valid) approaches. Ask a few students to put different approaches on the board.

SUMMARIZE

Expect to spend about 5 minutes.

7. When most students have gotten through worksheet step 11, reconvene the class. Focus attention on the board. *What do you think of these*

 **ACTIVITY NOTES**

approaches to solving the inequality? Are they valid? Solicit questions and perhaps other ideas.

8. ***Do the rules for solving an inequality make sense to you?*** For multiplying by a negative number, you might consider a specific example, such as multiplying both sides of the inequality $3 < 4$ by -1. ***Is it true that $-3 < -4$?*** You may get some disagreement on this question. As needed, remind students that *less than* means farther to the left on the number line, not necessarily closer to 0.

9. You might show how to manipulate the equation for when a lever will balance (presented before worksheet step 4). Multiplying and dividing both sides appropriately yields this proportion.

$$\frac{force}{load} = \frac{distance\ of\ load\ from\ fulcrum}{distance\ of\ force\ from\ fulcrum}$$

Do these ratios agree with your results? Help the class conjecture that they could have used this identity to find the tipping point.

10. ***What have you learned about levers and inequalities?*** As needed, help the class bring out at least the following two points.

   • Two ends of a lever are balanced if the products of their weights and the distances of the weights from the fulcrum are equal.

   • An inequality can be solved by balancing, like an equation, except that when multiplying or dividing by a negative the direction of the inequality is reversed.

11. ***What questions might you ask about levers?*** Encourage all inquiry.

   Does the height of the fulcrum (the log) make a difference?

   Is it possible to lift every object if you have a long enough lever?

ANSWERS

1. Answers will vary.

2. The rock is not lifted.

3. The rock lifts if the fulcrum is 1.6 feet from the load (6.4 feet from the force). The board will remain balanced at this point. If the fulcrum is less than 1.6 feet from the load (more than 6.4 feet from the force), the force will push the left end of the lever to the ground.

 ACTIVITY NOTES

4. Any letter is acceptable. These solutions use F.

5. $8 - F$

6. With strict inequalities, correct answers are $200F > 800(8 - F)$ or $800(8 - F) < 200F$. Students who believe that achieving balance is lifting the rock should use inclusive inequality instead of strict inequality.

7. $200F > 6400 - 800F$

8. $1000F > 6400$

9. $F > 6.4$

10. The fulcrum's distance from the force must be more than 6.4 feet.

11. This should be consistent with the Sketchpad result.

12. There are several alternatives. For example:

$$200F > 800(8 - F)$$
$$F > 4(8 - F)$$
$$F > 32 - 4F$$
$$5F > 32$$
$$F > 6.4$$

$$200F > 800(8 - F)$$
$$0.25F > 8 - F$$
$$1.25F > 8$$
$$F > 6.4$$

$$200F > 800(8 - F)$$
$$200F > 6400 - 800F$$
$$-6400 > -1000F$$
$$6.4 < F$$

13. Answers will vary. Sample answers: A 1-foot log might not fit within 1.6 feet of the center of a large rock. The left end of the board may be too high for the boys to stand on comfortably. The rock might slip off the board.

 ACTIVITY NOTES

14. Answers will vary. For example, the boys in the problem might think that the balance point will be farther from the rock. Or they might believe that the left end of a longer board will be lower than the left end of the shorter board. Less force will be required to lift the rock, because the distance from the fulcrum can be larger.

15.
$$200F > 800(12 - F)$$
$$200F > 9600 - 800F$$
$$1000F > 9600$$
$$F > 9.6$$

16. The balance point is 9.6 feet from the force (2.4 feet from the load). Closer than that to the load allows the force to push the lever to the ground.

Exploring Expressions and Equations in Grades 6–8 with The Geometer's Sketchpad
© 2012 Key Curriculum Press

Levers and Loads

Two boys stick one end of a long board under a large rock. One boy holds up the free end of the board while the other boy rolls a log under the board. Then they both climb up to the end of the board that's in the air to see whether they can lift the rock.

EXPLORE

1. The boys together weigh about 200 pounds and the rock weighs about 800 pounds. Do you think it's possible for them to lift the rock?

2. Open **Levers and Loads.gsp** and go to page "Lever." You can change the weights by dragging the top ends of the vertical line segments on the lower figure. Make the load 800 and the force 200 and press *Lift*. What happens?

3. Can you move the fulcrum so that the rock rises when you press *Lift?* If so, where would you put it? If not, why not?

To calculate where the fulcrum can be placed to lift that rock, you need to know that the lever will balance when this equation is true.

> (force) × (distance of force from fulcrum) = (load) × (distance of load from fulcrum)

In this situation, the force is 200, the load is 800, and the two distances are unknown. To express the fact that they won't balance, you'll need an inequality.

4. Start with the fulcrum's distance from the force. Choose a letter for the variable to represent the distance of the boys (the force) from the log (the fulcrum). _____

5. Without using another variable, write an expression for the distance of the rock (the load) from the fulcrum.

6. Write the inequality that is true when the boys are able to lift the rock.

Your goal is to isolate your variable in step 6 as if you were solving an equation. Follow steps 7–9 and write the resulting inequality after each step.

7. Apply the distributive property to the inequality in step 6.

8. Add the same value to both sides of the inequality in step 7 and combine like terms.

9. Divide both sides of the inequality in step 8 to isolate your variable. (Remember that multiplying or dividing both sides of an inequality by a negative number reverses the direction of the inequality.)

10. What does this result tell you about where the log can be placed so that the boys are able to lift the rock?

11. How closely does this result agree with what you found using Sketchpad?

12. Use the same rules to solve the inequality in at least one other way.

13. Why might this solution be unrealistic?

EXPLORE MORE

14. The boys try using a board that is 12 feet long. Why might they have decided on a longer board?

15. Set up and solve an inequality to find where the fulcrum needs to be along a 12-foot board to lift the rock.

16. Change the length of the lever in the sketch to 12 feet by dragging its right endpoint in the lower figure. What locations of the fulcrum lift the rock?

Shady Solutions:
Graphing Inequalities on a Number Line

 ACTIVITY NOTES

INTRODUCE

Project the sketch for viewing by the class. Expect to spend about 10 minutes.

1. Open **Shady Solutions.gsp.** Go to page "Number Line." Enlarge the document window so it fills most of the screen.

2. Drag point *x* back and forth on the number line. Ask, ***How would I graph the inequality x ≥ 2?*** Students should respond that you need to shade the portion of the number line to the right of 2, including 2.

3. Explain what you're doing as you demonstrate. ***To shade the number line in this Sketchpad model, I position point x at 2, select the short vertical segment through x, and choose*** Display | Trace Segment. ***I use the arrow key to move the segment to the right and shade that part of the number line.*** Explain that all the shaded values, such as 2, 5, and π, satisfy the inequality *x* ≥ 2, so they're part of the graph, whereas none of the unshaded numbers, such as 0, −3, and −10, satisfy the inequality, so they're not part of the graph. You may need to clarify the meaning of the word *satisfy* (to make the statement true). You might mention that when students make a graph like this on paper, they would put a filled-in dot at 2. Because the little segment in this Sketchpad model makes it easier to be precise, we are not using the dot.

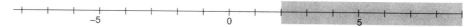

4. Go to page "A" and explain, ***In problem A there is a little segment that marks the expression x − 2 in addition to the segment that marks point x. Watch what x − 2 does while I drag x. Now we want to graph the slightly more complicated inequality x − 2 ≥ 5. We're still going to use the x-segment to shade values, but the relationship we're looking for is between x − 2 and 5. How should those markers be located? As I move x, tell me when the inequality is satisfied. You can all call out together. Start with x − 2 to the left of 5 and move x to the left.*** Students should call out for you to move the other way. Move right slowly. Students may start calling out when *x* reaches 5. Whether they do or not, build suspense at this point by saying, ***Is x − 2 greater than or equal to 5 yet?*** [No] Students should start calling out when *x* − 2 reaches 5.

5. Explain as you demonstrate. ***I start by moving x so that x − 2 lines up with 5. Then I select the x-segment and choose*** Display | Trace

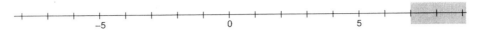
ACTIVITY NOTES

Segment. *Now, which way do I want to move x?* [Right] *I use the right arrow key to move x and shade that portion of the number line.*

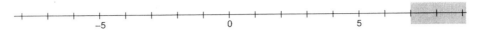

6. Explain, *The shaded part represents all the solutions to the inequality x − 2 ≥ 5. What simple inequality does this graph represent?* [x ≥ 7] *So the inequality x ≥ 7 is the solution to the inequality x − 2 ≥ 5.*

7. Tell students that they will graph the solutions to inequalities like this with increasingly complex expressions in them. *Using the graphs you'll be able to write simple solution inequalities like x is greater than or equal to or x is less than or equal to some number. You'll record the graphs and the solutions on your worksheet. If you finish the pages with the given inequalities, go on to the "Explore More" and the "Make Your Own" pages.*

DEVELOP

Expect students at computers to spend about 25 minutes.

8. Assign students to computers and tell them where to locate **Shady Solutions.gsp.** Distribute the worksheet. Tell students to work through step 14 and do the Explore More if they have time. Encourage students to ask their neighbors for help if they are having difficulty using the sketch.

9. Let pairs work at their own pace. As you circulate, here are some things to notice.

   • Watch for students who might be tracing the wrong segment. Students might be tempted to trace a segment representing an expression instead of the *x*-segment. Remind them that the solution consists of the *x*-values that make the inequality true.

   • Make sure students are finding the correct solution graphs and inequalities and recording them on their worksheets. If you see a mistake, ask the student to go to that page and show you how he or she found the solution.

   • In worksheet step 12, students should observe that the solution to the inequality in problem C has its inequality symbol reversed from the original. Students' explanations of why this happens will vary. Refer them back to worksheet step 8 and encourage them to give as much detail as they can. They should observe that the expression

Exploring Expressions and Equations in Grades 6–8 with The Geometer's Sketchpad

$-x + 2$ travels in the opposite direction from x because the expression includes the opposite of x.

- In worksheet step 13, students are asked to solve the inequalities algebraically. This activity is designed to introduce inequalities to students who have some experience with solving linear equations. For those students, as needed, suggest that they apply their equation-solving skills. You may need to make modifications, depending on the background of your students.

SUMMARIZE

Expect to spend about 10 minutes.

10. Gather the class. Students should have their worksheets with them. Begin the discussion by asking students to share their solutions to the inequalities.

11. Ask students if any strategies occurred to them for solving inequalities without Sketchpad. Whether or not they completed worksheet steps 13 and 14, students who compare their solutions to the original inequalities might note that the same strategies for solving equations (adding, subtracting, and dividing both sides of an equation) can also apply to inequalities.

12. Discuss worksheet step 12. Make sure students notice that the inequality symbol changed direction because of the $-x$. Depending on the background of your students, this observation can serve as a preview, introduction, or review of how to solve inequalities that involve multiplying or dividing by a negative number.

13. If time permits, discuss the Explore More and let students share inequalities they invented in the "Make Your Own" pages.

ANSWERS

4. Answers will vary. Students should observe that the marker for $x - 2$ moves in the same direction and speed as the marker for x and that the two markers are always two units apart.

5. $x \geq 7$

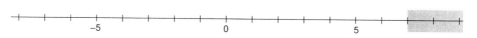

6. Answers will vary. Students should observe that the marker for $2x - 3$ moves faster than the marker for $x - 2$.

7. $x \geq 4$

8. Answers will vary. Students should observe that the marker for $-x + 2$ moves in the opposite direction from the marker for x.

9. $x \leq -3$

10. Answers will vary. Students should observe that the both the red and green markers move as x moves; on previous pages the green marker did not move, because it represented a constant value.

11. $x \leq 5$

12. Problem C. The symbol in the original inequality, $-x + 2 \geq 5$, is switched in the solution, $x \leq -3$. Explanations will vary. Students should observe that the marker for the expression $-x + 2$ moves in the opposite direction as the marker for x. They may explain that this happens because x and $-x$ are opposites of one another.

13. A. $x \geq 7$
    B. $x \geq 4$
    C. $x \leq -3$
    D. $x \leq 5$

14. The solutions should match. For inequalities like problem C, you must reverse the inequality when multiplying (or dividing) both sides by a negative number.

15. $-1 \leq x \leq 5$

To solve this algebraically, split the compound inequality into two separate inequalities, $-5 \leq 2x - 3$ and $2x - 3 \leq 7$, solve them, and combine their solutions.

16. Answers will vary. Sample answer: $-2x + 1 \geq 9$ (solution is $x \leq -4$)

17. Answers will vary. Sample answers: $2x + 1 \leq 2x$ (no solutions),
    $2x + 1 \geq 2x$ (entire number line)

Exploring Expressions and Equations in Grades 6–8 with The Geometer's Sketchpad

Shady Solutions

An inequality can have many solutions. You can represent them by shading the part of the number line that includes all the numbers that make the inequality true.

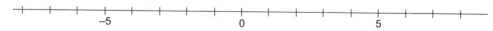

For example, the graph above shows the inequality $x \geq 2$. The shaded part includes all the numbers that make the inequality true.

What about a more complicated inequality, like $3x - 5 \leq 2x$? What values of x make it true? In this activity you'll explore graphs of solutions to inequalities like these and compare them to algebraic methods for finding these solutions.

EXPLORE

1. Open **Shady Solutions.gsp** and go to page "Number Line." Drag point x so that the inequality $x \geq 2$ is true.

2. Select the blue vertical segment through x and choose **Display | Trace Segment.**

3. Using the arrow keys to move x, shade the values of x on the number line that make the inequality $x \geq 2$ true. Your graph should end up looking like the example above.

4. Go to page "A." Drag x and observe how $x - 2$ behaves. How would you describe this behavior?

5. Trace the vertical segment through x and use the arrow keys to shade the solution to $x - 2 \geq 5$. Record the graph of the solution on the number line here.

Write the solution by circling the correct inequality symbol and filling in the blank with a number.

$x \leq \geq$ _____

6. Go to page "B." Drag *x* and observe how $2x - 3$ behaves. How does this behavior differ from that of $x - 2$ on page "A"?

7. Repeat the process described in step 5 to shade the solution to $2x - 3 \geq 5$. Record your graph and solution here.

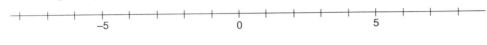

$x \leq \geq$ _____

8. Go to page "C." Drag *x* and observe how $-x + 2$ behaves. How does this behavior differ from those of pages "A" and "B"?

9. Find the solution to $-x + 2 \geq 5$. Record your graph and solution here.

$x \leq \geq$ _____

10. Go to page "D." Drag *x* and observe what happens. How does this behavior differ from those of the pages "A" through "C"?

11. Find the solution to $3x - 5 \leq 2x$. This time the variable apprears on both sides of the inequality. Record your graph and solution here.

$x \leq \geq$ _____

12. In one of these four problems, the inequality symbol in the solution reversed direction. In which inequality did this happen? How is this related to the behavior of that expression?

Exploring Expressions and Equations in Grades 6–8 with The Geometer's Sketchpad
© 2012 Key Curriculum Press

13. Use algebraic techniques to solve each of these inequalities.

   A. $x - 2 \geq 5$

   B. $2x - 3 \geq 5$

   C. $-x + 2 \geq 5$

   D. $3x - 5 \leq 2x$

14. How do your solutions to step 13 compare to those in steps 5, 7, 9, and 11? What additional rule might you need to solve some inequalities?

EXPLORE MORE

15. Go to page "Explore More." Use the sketch to find the solution to the compound inequality $-5 \leq 2x - 3 \leq 7$. Then explain how you could solve this problem using algebraic techniques.

16. Go to page "Make Your Own." Double-click the red expression, green value, and both question marks to change them. Record the inequalities you invent, their solutions, and the graphs of the solutions. Try to invent an inequality in which the inequality symbol of the solution is reversed from the original.

17. Go to page "Make Your Own 2." Double-click both expressions to invent inequalities in which the variable appears on both sides of the inequality symbol. Record the inequalities you invent, their solutions, and the graphs of the solutions. See whether you can write an inequality that has no solutions or whose solution is the entire number line.

In or Out: Graphing Linear Inequalities

INTRODUCE

Project the sketch for viewing by the class. Expect to spend about 5 minutes.

1. Open **In or Out.gsp** and go to page "Trace." Enlarge the document window so it fills most of the screen.

2. Explain, *Today you're going to work with inequalities in Sketchpad. You'll be able to use some of the same techniques you used in solving linear equations, but there will also be some important differences. The most important is that solutions to inequalities are represented by regions in the plane rather than by lines or curves. Before you begin, I'll demonstrate how the sketch works.*

3. Go to page "Trace" and begin dragging point A. Explain, *As you can see, for some locations on the grid, point A leaves a trace, but for others it does not. Look at the values of x, y, and x − 3 on the sketch and tell me how you think they are related to the tracing behavior of A.* Wait for students to provide conjectures.

4. At this point, you might want to stop with the introduction and let student pairs continue on their own. If you feel your students need more guidance, bring out the idea that A will leave a trace only when y is greater than x − 3. Then say, *Now predict a location for A that will leave no trace.* Students should be able to designate the bottom-right side of the coordinate system just by looking. More specifically, A doesn't leave a trace when it is on the bottom-right side of the line y = x − 3. *Because A traces only when its y-value is greater than its x-value minus 3, we say that the region we have colored with the trace is the solution to the inequality y > x − 3.*

DEVELOP

Expect students at computers to spend about 30 minutes.

5. Assign students to computers and tell them where to locate **In or Out.gsp.** Distribute the worksheet. Tell students to work through step 6.

6. As you circulate, here are some things to notice.

   • In worksheet step 2, students must look at specific points on the coordinate grid, determine their coordinates, and evaluate the expression x − 3. Students should notice that for the three points that are in the purple region, the value of x − 3 is less than y. Help students understand that the term x in the expression x − 3 refers to the x-coordinate of point A.

 ACTIVITY NOTES

- In worksheet step 3, invite students to use words such as "greater than" or "less than." They may also use the symbols $>$ and $<$.

- If possible, ask students to try equations other than $x - 3$ and $x + 1$. Consider, for example, expressions such as $2x$, $2 - x$, or $\frac{x}{2} + 5$.

- You might need to clarify for students that worksheet step 5 does not mean that *only* points in the first quadrant will leave traces.

- For worksheet step 6, several equations are possible. Challenge students to find at least three different solutions.

7. Gather the students together to compare their answers for worksheet step 6. Try a few of their examples to illustrate whether they work.

8. Have students continue. If you feel they need some guidance, go to page "Regions" and explain, **On this page Sketchpad can create the regions of inequality for us. Here the red line represents the function $f(x) = 1 - x$.** Press *Show region where $y > f(x)$*, and then press *Show region where $y < f(x)$*. Ask students to explain how the coordinates of A can help them figure out which region is less than $f(x)$ and which region is greater. Drag A to different locations.

9. Tell students to work through step 9 and do the Explore More if they have time.

SUMMARIZE

Project the sketch. Expect to spend about 10 minutes.

10. Bring students together and discuss their results for worksheet step 9. Use page "Regions" to construct the graph of the function $f(x) = 3 - 2x$ (or any other function the students have not yet explored). Double-click $f(x)$ on the top left of the screen to change it. Ask students how they would find the region $y < 3 - 2x$. Students may have different strategies. If students do not suggest it, point out that testing points on the axes such as $(0, 0)$ can make their computations easier. Emphasize the idea that they need to test one point in the plane (as long as the point is not on the line $y = f(x)$).

11. If students have not already noticed this, show them that it's also possible to infer which region is $<$ and which region is $>$ by a more visual approach. A line divides the plane into two parts, and the "less than" part is always the one in the bottom right or bottom left of the graph area.

EXTEND

You may want to introduce systems of inequalities. Open a new sketch and plot two linear functions such as $f(x) = x + 1$ and $f(x) = 3 - x$. Ask students to consider the inequalities associated with these functions and to write the systems of inequalities for which $(0, 0)$ is a solution. Repeat with other examples.

ANSWERS

2. Answers will vary.

3. When A leaves a trace, the y-coordinate is greater than $x - 3$. When A leaves no trace, the y-coordinate is less than $x - 3$.

4. When the y-coordinate is greater than the calculation $x + 1$.

5. Answers will vary. Possibilities include $y > -x$, $y > -5x$, and $y > -2x - 3$.

6. Answers will vary. Possibilities include $y > 4x$, $y > 10x$, and $y > 5x - 1$.

8. When A is in the yellow region, then its y-coordinate is *less than* $1 - x$ (the x-coordinate of A).

9. Use Sketchpad to create all the appropriate graphs. Answers for the Solution column will vary. For the third row, the function is $f(x) = 2x + 1$ and the inequality is $y > 2x + 1$. For the fourth row, the inequality is $y > x - 4$. For the fifth row, the function is $f(x) = \frac{x}{2}$.

10. Answers will vary. The quadratic inequalities will have either the inside or the outside of the parabola shaded.

In or Out?

In this activity you'll explore what it means to find a solution of an inequality and also how inequalities are solved using regions instead of lines or curves.

EXPLORE

1. Open **In or Out.gsp** and go to page "Trace."

2. Drag point A. Find three locations where A leaves a trace and three locations for which A leaves no trace. Fill in the table.

	$x =$	$y =$	$x - 3 =$
Trace	$x =$	$y =$	$x - 3 =$
	$x =$	$y =$	$x - 3 =$
	$x =$	$y =$	$x - 3 =$
No Trace	$x =$	$y =$	$x - 3 =$
	$x =$	$y =$	$x - 3 =$

3. Describe what happens to the values in each of the two regions (trace or no trace) using words or symbols.

   When A leaves a trace, the y-coordinate is _____ than $x - 3$.

   When A leaves no trace, the y-coordinate is _____ than $x - 3$.

4. Double-click the calculation $x - 3$, change it to $x + 1$ using the Calculator keypad, and click **OK**. Choose **Display | Erase Traces**. Drag A again. When does A leave a trace?

5. Erase the traces. Edit the calculation so that A always leaves a trace in the first quadrant. (It's fine if A also leaves a trace in other quadrants.) Write down the inequality.

6. Edit the calculation so that A leaves a trace when it's at $(-2, 3)$, but not when it's at $(1, 2)$, and write down the inequality.

7. Go to page "Regions." Drag *A* so that its coordinates are at about (3, 2). Press *Show region where y > f(x)*. You will see a green region that includes point *A*, which means *A* is a solution to the inequality $y > 1 - x$. Hide the region again.

8. Press *Show region where y < f(x)*. Drag *A* into the yellow region. Use the coordinates of *A* to explain why it satisfies this inequality.

9. Fill in this table. You can edit the function, drag point *A*, or show and hide the regions of inequality to help you out.

Function	Inequality	Example Solution	Graph
$f(x) = x + 4$	$y < x + 4$	(0, 0)	
$f(x) = 3 - x$	$y < 3 - x$		
$f(x) = x - 4$		(0, 4)	
	$y < \dfrac{x}{2}$		

Exploring Expressions and Equations in Grades 6–8 with The Geometer's Sketchpad

EXPLORE MORE

10. Go to page "Explore More" to investigate inequalities with quadratic and exponential functions. This page is similar to page "Trace."

    a. What does the graph of $y > x^2 - 3$ look like?

    b. Find a solution to $y < x^2 - 1$.

    c. Solve the inequality $y > -3x^2$ by sketching the graph of the region.

    d. Find a solution to $y > 3^x + 2$.

    e. Which inequality related to the function $f(x) = 2^x$ has the point $(3, 0)$ as a solution?

Around We Go: Equation of a Circle

INTRODUCE

Project the sketch for viewing by the class. Expect to spend about 5 minutes.

1. Open **Around We Go.gsp** and go to page "Coordinates."

2. Explain, *Today you're going to investigate the equation of a circle. You've been learning about many different types of equations, including linear and quadratic equations. The equation of a circle is quite different, as you will see. But you'll be able to use what you've learned about the Pythagorean Theorem to help you figure out how it works.* Review the Pythagorean Theorem as needed.

   Then continue, *To define an equation, we need to know how the coordinates of a point on the graph relate to each other. For the line $y = x + 3$, you know that for each point on the line, the y-coordinate will be 3 more than the x-coordinate. As I drag the point on the circle, you may find it difficult to see how its coordinates are related!* You might also review a different meaning of the equation of a line or curve: that a point with coordinates (x, y) lies on the line if, and only if, x and y satisfy the equation.

3. Drag the point around the circle and ask, *Does anyone see any special points where it looks as though the coordinates might be easier to see?* Students might mention the points that are on the axes, or perhaps grid points like $(3, 4)$ or $(4, 3)$ that go right through a grid point. *You will start by looking at some of the specific coordinates of various points on the circle and then see whether you can figure out how they are related.*

DEVELOP

Expect students at computers to spend about 30 minutes.

4. Assign students to computers and tell them where to locate **Around We Go.gsp.** Distribute the worksheet. Tell students to work through step 7.

5. Let pairs work at their own pace. As you circulate, here are some things to notice.

   • Make sure students are finding more than one possibility for each of the missing coordinates in worksheet steps 2a–2h. As needed, suggest that students drag point P to find other points with the same known coordinate.

   • In worksheet steps 3 and 4, help students plot the points to see whether they fall directly on the circumference of the circle. Tell them they can drag out the unit point to get a better idea of how well the plotted points fit the circle.

Exploring Expressions and Equations in Grades 6–8 with The Geometer's Sketchpad
© 2012 Key Curriculum Press

6. Gather students together to discuss their answers to worksheet step 7. Students may use counterexamples (as given in the solutions). They may argue that the equation $y = x + 5$ is constantly increasing as y increases, which is certainly not the case for the circle. Or they may point out that $y = x + 5$ is an equation of a line, which is not a circle. Ask students whether they have any other ideas about what the relationships might be. If necessary, draw attention to the two right angle triangles and ask students to recall what they know about right triangles.

7. Tell students to work through step 11 and do the Explore More if they have time. As you circulate, you may remind students of the Pythagorean Theorem, as needed.

SUMMARIZE

Project the sketch. Expect to spend about 10 minutes.

8. Gather the class. Students should have their worksheets with them. Use **Around We Go.gsp** to support the class discussion.

9. Go to page "Other Circles." Ask students to figure out what circle will go through the point $(6, 8)$. Students should be able to use the Pythagorean Theorem to find the radius of 10. Ask students to explain how they could argue that the circle of radius 10 is the only one that goes through the point $(6, 8)$ and is centered at the origin.

10. Ask students how the equation for a circle is different from the linear and quadratic equations they've worked with before. You might point out that, unlike a line and a parabola, each x-coordinate of a circle will have two values for its y-coordinate.

EXTEND

1. $(6, 8, 10)$ is called a Pythagorean triple. The three numbers are all whole numbers that satisfy the Pythagorean relationship. Another example is $(3, 4, 5)$. Challenge students to come up with other examples of Pythagorean triples.

2. **What other questions might you ask about equations of circles?** Encourage all inquiry. Here are some ideas students might suggest.

   Does the equation really apply to all points on the circle?

   What if you solve the equation for y?

Does every curve have an equation?

How would you graph that equation on a calculator?

Do three-dimensional shapes like spheres have equations?

ANSWERS

2. a. 5 or -5 b. 5 or -5 c. 3 or -3

   d. ≈ 4.6 or -4.6 e. 4 or -4.   f. ≈ 4.9 or -4.9

   g. 3 or -3 h. 4 or -4 i. 0

   j. 0

4. All except (d) and (f)

5. The y-coordinate begins increasing from 0 and reaches a maximum of 5 when the x-coordinate is 0, and then starts decreasing again.

6. No. That would not work when the point is at $(-5, 0)$ or when it's at $(3, 4)$, for example.

7. Answers will vary. Students should notice that the triangles remain right and that although they seem to vary in size, the same triangle can be found in each of the four quadrants. Students should also notice that each triangle has the same hypotenuse (the radius of the circle). They may also notice that as one leg of the triangle gets bigger, the other gets smaller.

8. $x^2 + y^2 = 25$

9. No, there are no other points on the coordinate grid that satisfy the equation. For all the other points, the sum of the squares will either exceed or be less than 25.

10. Answers will vary. For all of these points, $x^2 + y^2$ will be less than 25.

11. Radius 1: $x^2 + y^2 = 1$; radius 3: $x^2 + y^2 = 9$; radius 13: $x^2 + y^2 = 169$

12. The red number is the square of the radius.

Exploring Expressions and Equations in Grades 6–8 with The Geometer's Sketchpad

Around We Go

 Name:

In this activity you'll investigate the relationship between the *x*- and *y*-coordinates of points on a circle to see how they are related.

EXPLORE

1. Open **Around We Go.gsp** and go to page "Coordinates."

2. Drag the point on the circle to help you estimate the missing coordinate of each point on the circle—there may be more than one coordinate that works.

   a. (0,   )   b. (, 0)   c. (, 4)   d. (−2,   )   e. (3,   )

   f. (, 1)   g. (−4,   )   h. (, −3)   i. (, −5)   j. (5,   )

3. Use **Graph | Plot Points** to plot the coordinates you found in step 2. Use the tab to switch between coordinates and press **Plot** after each point. When you finish plotting points, press **Done**.

4. Which of the coordinates in step 2 are exactly on the circle?

5. Press *Trace Circle*. The point will move from (5, 0) once around the circle counterclockwise. How does the *y*-coordinate of the point change as the *x*-coordinate approaches zero and becomes negative?

6. Is the relationship between *x*, *y*, and the radius (which is 5) represented by the equation $x + y = 5$? Explain.

7. Press *Show Right Triangle*. The new segments show that you can make a right triangle. Drag the point around the circle and describe everything you notice. What changes? What stays the same?

8. Use what you know about the relationship among the side lengths of a right triangle to write an equation that relates the *x*-coordinate, the *y*-coordinate, and the radius of 5.

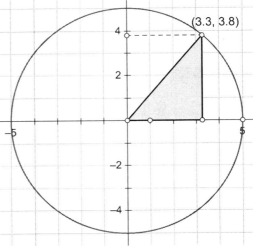

9. Does any point *not* on the circle satisfy the equation? Explain.

10. Give the coordinates of three points in the interior of the circle. What can you say about the *x*- and *y*-coordinates of points inside the circle?

11. Go to page "Other Circles." Drag point *A* to investigate circles with smaller and larger radii. How can you change the equation from worksheet step 8 to represent a circle with a radius of 1, 3, or 13 units?

EXPLORE MORE

12. Go to page "Explore More." The equation of a circle shown on the sketch is given as $x^2 + y^2 = 4$. Drag the point to different locations of the circle and verify that this equation does indeed hold, using the point's *x*- and *y*-coordinates. Then change the value of the red number 4 by double-clicking it. Describe the relationship you see between this value and the radius of the circle.

Exploring Expressions and Equations in Grades 6–8 with The Geometer's Sketchpad
© 2012 Key Curriculum Press

Make It Move: Distance-Rate-Time Animations

INTRODUCE

Project the sketch for viewing by the class. Expect to spend 15 minutes.

1. Explain that Sketchpad animation can model distance-rate-time problems. Say, *To see how, we will solve a problem together and build a Sketchpad model.* Open **Make It Move Present.gsp** to page "Car and Bike Problem." Read the problem, recording the distance and rates on the board.

   A car and a bicycle that are 15 miles apart start moving toward each other at the same time. The car travels 40 mi/h, and the bicycle travels 20 mi/h. How far are they from their starting points when they meet?

2. *A good problem-solving technique is to act out the problem or draw a picture. First, let's draw a picture to represent this situation.* Have students draw their own diagrams, and ask a student to draw one on the board. Don't use Sketchpad at this point.

3. *How could you solve the problem?* Students might suggest making a table and using a guess-and-check strategy. They might use measurement units to guide arithmetic calculations. Or they might write and solve an equation or proportion. Students who have a good understanding of the proportional nature of motion problems may observe that because the speeds are in a 2:1 ratio, the distances will be as well. Two distances in a 2:1 ratio that add to 15 miles are 10 miles and 5 miles. Other approaches will lead to the same result for when the vehicles meet; the car is 10 miles from where it started and the bike is 5 miles from where it started.

4. *We can act out this solution using animation in Sketchpad. The animation can help us see the action and test our solution.* Begin demonstrating how to create the animation. You might ask students to follow along in their own new sketches.

Point out that the directions are on the worksheet, for students to follow when they work on their own.

   - Open the presentation sketch to the blank page. Maximize the sketch window.

   - Draw a segment. With only the endpoints selected, choose **Measure | Distance.**

   - With the **Arrow** tool, drag either endpoint to make the distance between them about 15 cm. (You can also use the arrow keys on your keyboard to do this. You may not be able to get exactly 15 cm, but get within 1 or 2 hundredths.)

5. Ask, **What does the distance AB represent?** [The distance represents the initial distance between the car and the bike. In this model, the scale is 1 cm : 1 mi.] If some students are unsure, refer them to the drawing on the board and their own work.

Explain that when students make their own models, they can choose scales that will allow the models to fill the screen as much as possible. To model how to change the scale, choose **Edit | Preferences**. Look for Distance in the pop-up menu under the Units tab, change **cm** to **pixels,** click **OK,** and look at the units on the measurements. You might do the same after changing the distance units to inches. Change the units back to centimeters to proceed with your current model.

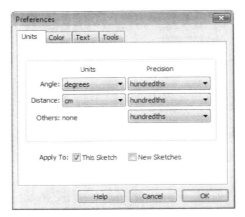

6. Continue demonstrating.

   - Construct a point *C* on the segment.

   - Select points *A* and *C* and choose **Measure | Distance.**

   - Drag point *C* so that *AC* is as close as possible to 12 cm.

   - Ask, **What does point C represent? Why is it located here? Look at your sketch.** [The car and bike meet 12 miles from the car's starting point. Point *C* represents the point where they meet.]

7. **Now we'll add the car and the bike.**

   - Construct two more points on the segment, one on each side of point *C.*

Exploring Expressions and Equations in Grades 6–8 with The Geometer's Sketchpad
© 2012 Key Curriculum Press

- Using the **Text** tool, double-click on the point on the left and change the label to *Car*. Change the other point's label to *Bike*. Your model should look like this.

AB = 15.0 cm   CA = 10.0 cm
Car   Bike
A   C   B

You may wish to post the directions for getting clip art images or accessing students' digital photos.

- If students are using **Make It Move.gsp,** they'll copy from page "Art." For other art, explain how to get the images.

- Copy the car art or a student's digital photo, then select point *Car* in your sketch and choose **Edit | Paste Picture.** Repeat to paste the bike art or a photo to point *Bike*.

8. *Now let's tell Sketchpad how we want these vehicles to move.*

- Select the *Car* and *Bike* points and choose **Edit | Action Buttons | Animation.**

- In the dialog box that appears, set the Car direction to **forward.** Choose **other** for speed and enter 40.

- Click on the Bike animation statement. Set it to **backward.** Set the speed to 20. Click **OK.**

9. *Let's get ready to animate.* Invite a volunteer to the computer to carry out the next steps as you explain what to do.

- *Move points Car and Bike to their starting positions.* Let students agree that these are at the segment endpoints.

- *Don't click the button yet! When you press* **Animate,** *the points will move toward each other, and when you press the button again, they'll stop. You want to be ready to stop the animation exactly as they meet to see whether they meet at point C, as you have predicted.*

> · *Ready? Press* **Animate Points.**
>
> · *Oops! Too fast. We can adjust the speed in Sketchpad.*

10. Have the volunteer change the speed, following your directions.

> · *Select the* **Animate Points** *button and choose* Edit | Properties. *Set the Car speed to 4 and the Bike speed to 2.*
>
> · *Move the Car and Bike back to their places. Press* **Animate Points** *again.*
>
> · *You should be able to demonstrate that they meet at point C.*

11. *In the problem, the car's speed was 40 mi/h and the bike's speed was 20 mi/h. Why would animation speeds of 4 and 2 work in our model?*

    The discussion should elicit the idea that the ratio of speeds must be in the correct proportion. Because 40:20 = 4:2, the speeds 4 and 2 will work. *Would other speeds work to slow it down even more?* [Yes, 2 and 1 or 0.4 and 0.2] *Why?*

    Point out that the animation speeds don't have units. So, the 40 and 20 used initially aren't the actual speeds of the car and bike in the problem; they can be used to represent the speeds because they are in the correct proportion.

    What if the car were going 50 and the bike were still going 20? Would point C move to the left or the right? Why? [It would move closer to the bike's origin, because the car would travel farther before meeting the bike.]

> To select the *Animate Points* button, click on the "handle," the black strip at the left side of the button. Do not simply click on the button.

DEVELOP

> Expect students at computers to spend 60 minutes.

12. Distribute the worksheet, assign students to computers, and ask them to open a blank sketch. Show how they can choose **File | Document Options** to add and name a new page to the document for each model they create.

    When you have trouble with Sketchpad, ask other students to help you.

13. Direct students to work in pairs to solve and then create animations for problems 1–3 on the worksheet. Emphasize that students should draw a picture and solve each problem on paper before they model it with Sketchpad.

 ACTIVITY NOTES

As students work, you can assess their understanding of distance-rate-time problems.

14. Suggest that students choose scales for their Sketchpad models that will allow their models to fill the screen as much as possible. Students can change Preferences in Sketchpad to show distances in inches or pixels, if that makes it easier.

15. As you circulate, observe the strategies students are using in the paper-and-pencil solutions to the problems. Encourage them to include diagrams in their solutions to help them when they create the Sketchpad models.

    Remind students that they can choose **Edit | Preferences** to choose a scale for their Sketchpad models that will allow them to fill the screen.

16. As students create their Sketchpad models, ask questions to draw out their thinking about the animation speeds they are using.

    Tell me about the animation speeds you've been trying. What have you found out?

    Are there other speeds that will work?

    Could you slow the animation down even more?

17. Be prepared to help students who want to create *Return to Start* buttons. The best way to create the starting points for the antelope and the teacher is to translate point *A* to the right the appropriate distance. Then the point will not move relative to point *A*.

18. Have students save their sketches, preferably in a place that will be accessible to the computer connected to the overhead display so that they can share their work in the summary discussion.

SUMMARIZE

Expect to spend 15 minutes. To facilitate discussion, invite students to show their models using a computer connected to a projection device, if possible.

19. Invite students to share their experiences solving and modeling the problems. Some students may have run into errors in their solutions when they set their models in motion. Let the mistakes, corrections, and successes be the focus of this class discussion. Ask questions such as these.

    As you made and used your models, did anything surprise you?

    Did you create a model that didn't work at first?

    What did you do to make it work?

    Did making the models help you think about this type of problem? In what ways?

Was one problem harder for you to solve than others? What did you learn about solving it?

20. Make sure students have the opportunity to share their solutions to any problems they invented.

21. If time allows, ask these questions or present them as writing prompts.

How would you describe to someone else one of the problems and models you worked with today?

How could you use animation to help you solve a distance-rate-time problem?

ANSWERS

Problem 1: Point *C* is at Elko, where the cars will meet. The distance from *A* (Reno) to *B* (Salt Lake City) is 500 units, and from *A* to *C* is 280 units. The car that traveled from Reno went 280 miles, traveling 70 mi/h, so it traveled for 4 hours. The other car, therefore, must travel at a rate of 220 mi/4 h, or 55 mi/h.

Problem 2: The distance from *A* (the cheetah) to *B* (the antelope) is 100 m and from *A* to *C* is 500 m. The antelope ran 500 m − 100 m, or 400 m in 10 sec, which is 40 m/sec.

Problem 3: *C* is the point where the car catches up with the teacher on her bike. The husband found the lunch in 12 minutes $\left(\frac{1}{5}\,h\right)$, so the teacher is 3.6 mi from home when he leaves in the car. Her husband is catching up to her at 45 − 18, or 27 mi/h. To close the 3.6 mi will take $\frac{3.6\ \text{mi}}{27\ \text{mi/h}} = \frac{2}{15}\,h = 8$ min. In that time, the husband goes 6 mi and the wife, another 2.4 mi.

Make It Move

In this activity you'll solve distance-rate-time problems. Then you will model your solutions using animation in Sketchpad.

Problem 1: Elko is between Reno and Salt Lake City. From Reno to Elko is about 280 miles and from Salt Lake City to Elko is about 220 miles. One car leaves Reno and travels east toward Elko at 70 mi/h. If the other car leaves Salt Lake City at the same time, how fast should it travel to arrive in Elko at the same time as the car from Reno?

Problem 2: A cheetah starts chasing an antelope 100 meters away and at the same instant the antelope starts running away. The cheetah catches the antelope in 10 seconds, after running 500 meters. How fast did the antelope run?

Problem 3: A teacher left for school on her bike, traveling at 18 mi/h. Twelve minutes after she left, her husband discovered she had forgotten her lunch. He drove his car at 45 mi/h to catch up to her. How far from home will they be when he catches up to her?

1. On paper, draw a picture of each problem.

2. Solve each problem using any method you know.

3. In a new sketch, you'll make a separate page for each problem. Choose **File | Document Options**. In the dialog box pop-up menu, choose **Add Page | Blank Page**. You can give the page a name or leave it as "2." Add another blank page before you close the dialog box.

 4. On the first page, construct a segment.

 5. Select the segment's endpoints and choose **Measure | Distance**.

6. Drag either endpoint to make the distance between the endpoints the same as the distance between the starting points of the problem (using a scale factor).

 7. For Problem 1, construct a point C on the segment. For Problems 2 and 3, use the **Ray** tool to extend $\overline{AB}$ and construct point C on the ray.

 8. Select points A and C and choose **Measure | Distance**. Now drag point C so that AC is the distance from one starting point to the place where the two vehicles or animals meet.

9. For each problem, describe what point *C* represents. What determined where you located *C*?

10. Construct and label two more points on the segment to represent the vehicles or animals.

11. If you want to use art, copy one image from **Make It Move.gsp** or another file. Select a point in your sketch and choose **Edit | Paste Picture.** Repeat to paste an image to the other point.

12. Select the points that you will animate. Choose **Edit | Action Buttons | Animation.** In the dialog box, set direction and speed for each point.

13. Move the points to their starting places.

14. Press *Animate Points.*

15. Too fast? How might we fix that? Sketchpad can slow the animation down. Select the handle of the *Animate Points* button on the left side without pressing the button, and choose **Edit | Properties.** Change the speeds.

16. Move the vehicles or animals back to their starting points and try the *Animate Points* button again. You should be able to demonstrate that they meet at point *C*.

17. Do the points meet at *C*? Explain your results. If you need to change your sketch, tell what you will change. Then report the results when you try again.

EXPLORE MORE

18. You can make a button that will return each object to its starting place. For Problem 1, select in order the points for Car 1, Reno, Car 2, and Salt Lake City. Choose **Edit | Action Buttons | Movement** and set the speed to **instant.** You might change the label to *Return to Start* before you click **OK** to close the dialog box. For Problems 2 and 3, you will need to create a starting point for the antelope or teacher before you make your movement button.

19. Make your own distance-rate-time problem and solve it. Then add a blank page to your sketch document, create an animated model, and test your solution. Record your results and observations.

Exploring Expressions and Equations in Grades 6–8 with The Geometer's Sketchpad
© 2012 Key Curriculum Press

Polynomials

Tiling in a Frame: Multiplying Polynomials

Students use Sketchpad algebra tiles to multiply polynomials. Using the polynomial factors as dimensions, they build rectangles out of tiles. The area of the completed rectangle represents the product.

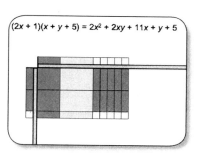

Tiling Rectangles: Factoring Polynomials

Students use Sketchpad algebra tiles to factor polynomials. They represent the polynomial with a collection of tiles and arrange them into a rectangle. The dimensions of the completed rectangle represent the factors.

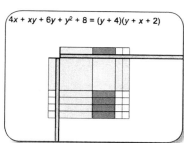

Dynamic Tiles: Evaluating Polynomial Expressions

Students use dynamic algebra tiles to evaluate polynomial expressions for different values of x and y by dragging sliders. Students fill in the areas with unit tiles and count the area, and then evaluate the expressions on paper to verify products of polynomials.

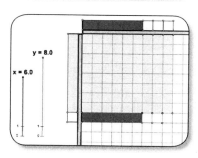

Rooting for Roots: Factoring and Graphing Quadratics

Students use a dynamic area model to explore the multiplication of binomials. Students also graph quadratic equations and relate the binomial products and the resulting graphs to the roots of the quadratic equations.

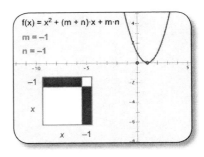

Exploring Expressions and Equations in Grades 6–8 with The Geometer's Sketchpad

Tiling in a Frame: Multiplying Polynomials  ACTIVITY NOTES

This activity can be used by students whether or not they have had hands-on experience with physical algebra tiles. Most students, especially tactile learners, benefit from manipulating physical tiles. This activity offers a similar model, using Sketchpad tiles to represent polynomials.

If your students have experience with physical tiles, you might have them use custom tools to construct arrangements that will update dynamically as students drag sliders for *x* and *y*. If so, students should use page "Explore More," rather than page "Frame," to complete the worksheet. Students who do not have experience with physical tiles will need to work with the Sketchpad tiles before they can effectively use the custom tools.

INTRODUCE

Project the sketch for viewing by the class. Expect to spend about 10 minutes.

1. The first page of the student worksheet—"Get Ready"—can be done in class the day before students use computers, assigned as homework, or used as a warm-up. If your students have never used algebra tiles, you should introduce them, as explained in the next step, *before* assigning Get Ready. In any case, you should review the Get Ready questions before starting the activity. Ask for volunteers to explain their answers. Emphasize the relationship between dimensions and area.

2. Open **Tiling in a Frame.gsp** and go to page "Tiles." Make sure that the dimensions are showing. Drag tiles next to other tiles to compare their dimensions. Explain that tiles are named after their areas and make sure students can identify all the names. Drag the *x* and *y* sliders to show that the area expressions match the dimensions for all values. Explain that this activity uses rigid tiles, but that students can explore dynamic tiles in the Explore More.

3. Go to page "Example 1" (or go to page "Frame" and make up your own examples). Enlarge the document window so it fills most of the screen. Explain, *Today you're going to use Sketchpad to create rectangles using tiles. First you'll drag tiles from the stacks to the outside edges of the frame. Here you can see x + 2 along the top and x + 3 along the side. These are the dimensions of the rectangle.* Students should pay attention to how these *dimension tiles* are represented on the frame—along the top and outside left edges of the frame, and separated by like terms.

4. Explain, *You'll drag tiles from the stacks into the frame to create rectangles. The tiles should be arranged so that there are no gaps or overlaps.* Press *Show Tile Area.* Explain that the dimensions of this rectangle match those along the outside of the frame. The "seams" in the completed rectangle create straight segments that extend the entire length or width of the rectangle and are aligned with the "seams" in the dimension tiles. Press *Show Blueprint Lines* if you want to emphasize this, but students will not have blueprint lines for the problems on the worksheet. You can show the blueprint lines and tile areas simultaneously to demonstrate that rays constructed from the dimension tiles overlap every seam without going through the interior of any of the tiles.

5. Ask, **What is the area of the completed rectangle?** Then use a text caption, or use the board, to write the multiplication problem.

$$length \times width = area$$

$$(\text{or, } base \times height = area)$$

$$(x + 2)(x + 3) = x^2 + 5x + 6$$

Ask what each polynomial represents. Make sure that students know their basic multiplication terminology and notation. They should know what *factors* are and that a product can also be symbolized using only parentheses (without the $\times$ symbol).

6. If you wish to give more guidance, go to page "Example 2." You might review the example together with the class, ask student pairs to review it on their own, or mention it to student pairs only if they need support.

DEVELOP

Expect students at computers to spend about 30 minutes.

7. Assign students to computers and tell them where to locate **Tiling in a Frame.gsp.** Distribute the worksheet. Tell students to work through step 14 and do the Explore More if they have time. Encourage students to ask their neighbors for help if they are having difficulty with Sketchpad.

8. Let pairs work at their own pace. As you circulate, here are some things to notice.

   • Make sure students are placing tiles so that they match both dimensions along the outside of the frame. In particular, make sure students are not using 3 unit tiles as the equivalent of an x tile.

- If a student pair is really struggling, you might show them how to construct blueprint lines by using the **Ray** tool to click on the intersections of the adjacent tiles along the outside of the frame.

- In worksheet steps 6–9, remind students to press *Reset* to return all of the tiles to the stacks after each problem.

SUMMARIZE

Project the sketch. Expect to spend about 5 minutes.

9. Gather the class. Students should have their worksheets with them. Ask, ***How did you multiply polynomials today?*** Establish as a class that multiplying polynomials can be represented by the multiplication equation *length* × *width* = *area* (or, *base* × *height* = *area*), where the dimensions correspond to the factors and the area corresponds to the product.

10. ***How did you decide which tiles to use to build a rectangle? How did you determine which tile goes in each place?*** Students should understand that every tile is a product of two dimension tiles, one on each side of the frame, and that they have to consider both dimensions when choosing their area tiles.

11. ***Are there shortcuts to multiplying? Do you always have to use tiles to multiply polynomials?*** Students should be able to make a multiplication table with dimensions that correspond to the factors and predict the quantity and type of tiles that belong in each region.

12. If time permits, discuss the Explore More. Let students show arrangements they made using the custom tools and how the arrangements adjust to the values of the sliders. You may consider giving all students the chance to work with the dynamic tile tools on another day.

EXTEND

1. Ask, ***What are some limitations of building rectangles with algebra tiles? Do the limitations still exist with multiplication tables?*** Limitations include the inability to represent negative values, nonlinear dimensions, or variables other than *x* and *y*, and the lack of enough pieces for large polynomials. Multiplication tables do not have these limitations.

2. Ask, *If the product of two polynomials can be represented by the area of a rectangle, how could you represent the product of three polynomials?* The product of three polynomials can be represented by the volume of a rectangular prism.

ANSWERS

Get Ready

A. $EF = 7$, $BC = 7$. The height of a rectangle is constant.

B. $2x^2 + xy + 3x + 2y + 3$

C. $2x + 2y + 4$; $4x + 4$; $4x + 2y + 6$

D. $2 \times 5 = 10$; $6 \times 7 = 42$; $a \times b = ab$

E.

```
   ┌──────────┐
 3 │          │
   │          │
   └──────────┘
        4
```

F.

```
      ┌────────┐
x + 1 │        │
      │        │
      └────────┘
        x + 2
```

Answers for area will vary. Students should communicate that they understand that $x + 1$ and $x + 2$ should be multiplied.

Explore

4.

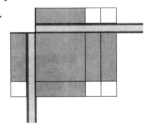

$(x + 1)(x + 2) = x^2 + 3x + 2$

6.

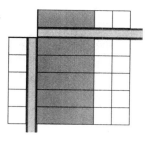

$$(5)(x + 2) = 5x + 10$$

7.

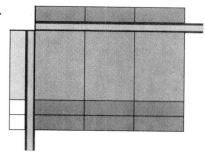

$$(y + 2)(3x) = 3xy + 6x$$

8.

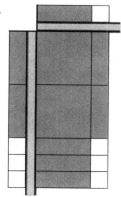

$$(2x + 3)(x + 1) = 2x^2 + 5x + 3$$

9.

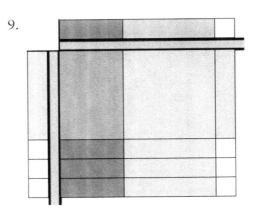

$$(y + 3)(x + y + 1) = xy + y^2 + 3x + 4y + 3$$

10. Building the rectangle would require 40 tiles. (6 x^2 tiles, 22 x tiles, and 12 unit tiles)

11. The red points appear wherever like terms are separated in the dimension tile.

12.

	$3x$	2
$2x$	$6x^2$	$4x$
6	$18x$	12

$$(2x + 6)(3x + 2) = 6x^2 + 22x + 12$$

13.

	x	y
$4x$	$4x^2$	$4xy$
7	$7x$	$7y$

$$(4x + 7)(x + y) = 4x^2 + 4xy + 7x + 7y$$

14.

	x	y	1
y	xy	y^2	y
3	$3x$	$3y$	3

$$(y + 3)(x + y + 1) = xy + y^2 + 3x + 4y + 3$$

The number of rows and columns correspond to the number of terms in each of the factors.

Exploring Expressions and Equations in Grades 6–8 with The Geometer's Sketchpad
© 2012 Key Curriculum Press

Tiling in a Frame

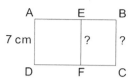 **Name:**

GET READY

A. In the figure below, *ABCD* is a rectangle. Segment *EF* is perpendicular to segment *DC*. What are the measures of segments *EF* and *BC*? Why?

```
   A      E      B
       ┌──────┬──────┐
7 cm   │      │?     │?
       └──────┴──────┘
   D      F      C
```

B. Write the following area collection as a sum. Combine like terms.

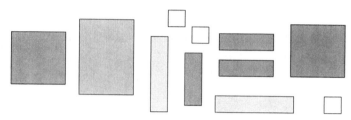

C. Find the perimeter of the following rectangles.

a.   b.   c.

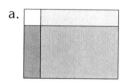 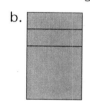 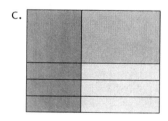

D. Find the area of the rectangles shown here. Write the area as a product of the dimensions.

```
                                6 ft
                           ┌──────────┐
        ┌──────────┐       │          │        ┌──────────┐
2 in    │          │  7 ft │          │   a    │          │
        └──────────┘       │          │        │          │
          5 in             └──────────┘        └──────────┘
                                                    b
```

____ ____ = ____   ____ ____ = ____   ____ ____ = ____

Dimensions   Area   Dimensions   Area   Dimensions   Area

E. Draw a picture of a rectangle representing the product $(3)(4) = 12$.

F. Draw a picture of a rectangle representing the product $(x + 1)(x + 2)$. What do you think this area will be?

Tiling in a Frame

 Name:

In this activity you'll explore the areas of rectangles when you know the dimensions, even if the length and width include variables.

EXPLORE

1. Open **Tiling in a Frame.gsp.** If you don't know the names of the tiles, go to page "Tiles." Otherwise, go to page "Frame."

2. Drag tiles from the stacks to the outside edges of the frame to represent the product $(x + 1)(x + 2)$.

3. Use tiles to build a rectangle inside the frame with dimensions that match those along the outside edges of the frame.

4. Make a sketch of the finished arrangement and write out the multiplication as an area equation.

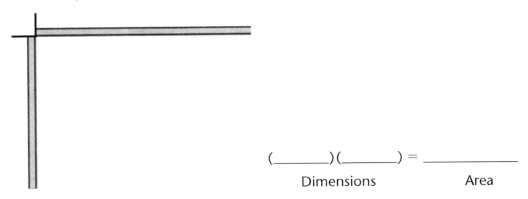

(_____)(_____) = _____

Dimensions   Area

5. Press *Reset* to return all tiles to the stacks.

6. Repeat steps 2–5 for the product $(5)(x + 2)$.

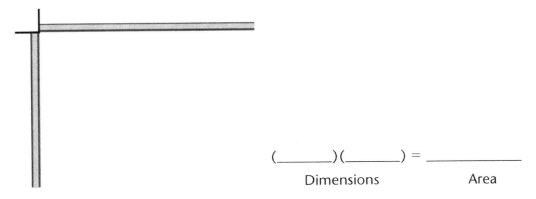

(_____)(_____) = _____

Dimensions   Area

Exploring Expressions and Equations in Grades 6–8 with The Geometer's Sketchpad
© 2012 Key Curriculum Press

7. Multiply $(y + 2)(3x)$. Sketch the finished arrangement.

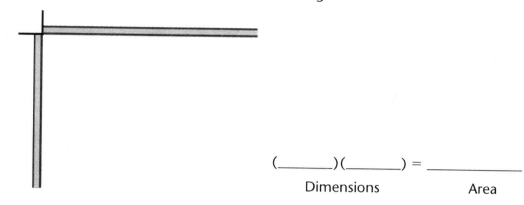

(_____)(_____) = _____

Dimensions   Area

8. Multiply $(2x + 3)(x + 1)$. Sketch the finished arrangement.

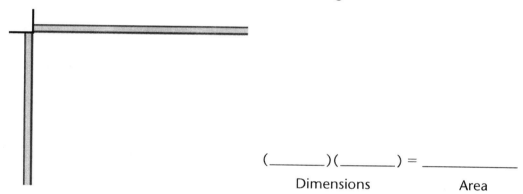

(_____)(_____) = _____

Dimensions   Area

9. Multiply $(y + 3)(x + y + 1)$. Sketch the finished arrangement.

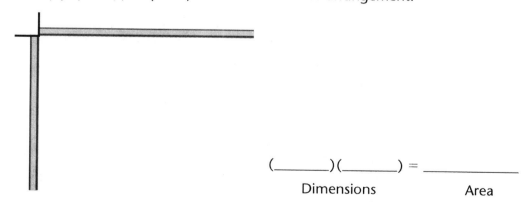

(_____)(_____) = _____

Dimensions   Area

10. Go to page "Regions." Suppose you had to multiply $(2x + 6)(3x + 2)$. How many tiles would you need to build a rectangle for this?

11. Press *Show Regions.* Look at the red points, where the dashed lines begin. How are these red points related to the dimension tiles?

12. Instead of building a rectangle, you'll divide the space into regions, separating like terms. Complete the multiplication table and write the multiplication equation.

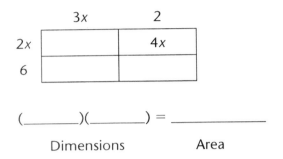

(_____)(_____) = _____

    Dimensions   Area

13. Multiply $(4x + 7)(x + y)$ as you did in step 12.

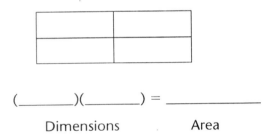

(_____)(_____) = _____

    Dimensions   Area

14. Multiply $(y + 3)(x + y + 1)$ using a multiplication table. How can you determine the number of columns and rows you need in the table?

EXPLORE MORE

15. Go to page "Explore More." Use the dynamic tile tools in the Custom Tools menu to construct an arrangement. Choose a tool and click an existing point in the sketch. There are three points you can start with, one for each dimension along the frame and one for the inside of the frame. To connect new tiles to existing tiles, move the pointer until a point on the existing tile becomes highlighted and then click.

Drag the *x* and *y* sliders to see how the arrangement adjusts to their new values. If you've constructed your arrangement correctly, it will hold together when you drag the sliders.

Exploring Expressions and Equations in Grades 6–8 with The Geometer's Sketchpad
© 2012 Key Curriculum Press

Tiling Rectangles: Factoring Polynomials  ACTIVITY NOTES

This activity can be used by students whether or not they have had hands-on experience with physical algebra tiles. However, students should have already completed the activity Tiling in a Frame, which introduces the Sketchpad algebra tiles.

If your students have experience with physical tiles, you might have them use custom tools to construct arrangements that will update dynamically as students drag the *x* and *y* sliders. If so, students should use page "Explore More," rather than page "Frame," to complete the worksheet. Students who do not have experience with physical tiles will need to work with the Sketchpad tiles before they can effectively use the custom tools.

INTRODUCE

Project the sketch for viewing by the class. Expect to spend about 10 minutes.

1. The first page of the student worksheet—Get Ready—can be done in class the day before students use computers, assigned as homework, or used as a warm-up. In any case, you should review the Get Ready questions before starting the activity. Ask for volunteers to explain their answers. Emphasize that the area of a rectangle can be expressed as the product of the dimensions and that the *x* and *y* tiles can be used to represent both area and dimension because they have a unit width.

2. Open **Tiling Rectangles.gsp** and go to page "Tiles." Remind students that tiles are named after their areas. Drag the *x* and *y* sliders to show that the area expressions match the dimensions for all values. Explain that this activity uses rigid tiles, but that students can explore dynamic tiles in the Explore More.

3. Go to page "Example 1" (or go to page "Frame" and make up your own examples). Enlarge the document window to fill the screen. Explain, *Today you're going to use Sketchpad to make rectangular arrangements with tiles, but this time you won't be given the dimensions. Instead you'll be given the area and you'll need to find the dimensions. You'll do this by making a rectangle in the frame. For example, here is one way to arrange the expression $4x + xy + 6y + y^2 + 8$ into a rectangle. Notice that the tiles are arranged so that there are no gaps or overlaps.* Press *Show Blueprint Lines*. Then press *Hide Area Tiles*. Explain that the blueprint lines also work in reverse. Students can check that the "seams" in the rectangle align with the "seams" along the outside of the frame.

4. Drag an inappropriate tile, such as an xy or a y^2, along the outside of the frame and say, ***It looks like this piece fits here. Is this correct?*** Students should understand that dimension tiles need to have a unit width, which in this model includes only the 1, x, and y tiles. The tiles along the outside of the frame should not extend beyond the black segments in the upper-left corner of the frame.

5. Press *Show Dimension Tiles*. ***What are the dimensions of the completed rectangle?*** $[(y + 4)$ and $(y + x + 2)]$ ***The dimensions represent the factors of the original polynomial. You can think of factoring as a multiplication problem in reverse.*** Use a text caption, or use the board, to write the multiplication problem.

$$area = length \times width$$

$$(or, area = base \times height)$$

$$4x + xy + 6y + y^2 + 8 = (y + 2)(y + x + 4)$$

6. Go to page "Example 2" and press *First Attempt*. ***This is my first attempt at making a rectangle for*** $x^2 + 7x + 10$. ***What do you think?*** Elicit the ideas that the tiles do not line up correctly and that the overall shape is not actually a rectangle. Press *Show Blueprint Lines* to show that some of the blueprint lines pass through the x^2 tile. Then press *Second Attempt*. You might clarify that some of the horizontal x tiles needed to be rotated to be vertical in order to make the rectangle, and that is fine. They are the same pieces and have the same area. ***Can you tell what the factors will be now?*** $[(x + 5)$ and $(x + 2)]$. Make sure students understand that the dimensions represent the factors of the polynomial.

DEVELOP

Expect students at computers to spend about 30 minutes.

7. Assign students to computers and tell them where to locate **Tiling Rectangles.gsp.** Distribute the worksheet. Tell students to work through step 15 and do the Explore More if they have time. Encourage students to ask their neighbors for help if they are having difficulty with Sketchpad.

8. Let pairs work at their own pace. As you circulate, here are some things to notice.

   • Make sure students follow the "no gaps or overlaps" rule when making their rectangles. The tiles are noncommensurate and a tile will not align unless it shares the same dimensions as the adjacent tile. You can

Exploring Expressions and Equations in Grades 6–8 with The Geometer's Sketchpad
© 2012 Key Curriculum Press

use the idea that blueprint lines must run through the rectangle to the frame uninterrupted.

- Check that students use only the 1, x, and y tiles to represent dimensions. The x^2, xy, and y^2 tiles can only represent area.

- Some students might need to be assured that horizontal and vertical versions of the same tile are in fact interchangeable. They have the same area. You might connect this to the conservation of area or the commutative property of multiplication.

- Although students can find many ways to make a rectangle, some students might find it easier to always group similar tiles together, with the larger tiles in the upper-left corner and the unit tiles in the lower-right corner. This facilitates trinomial factoring by showing the factors of the constant term as rectangular dimensions.

- Tiles can be dragged back to the stacks if a student decides that a piece is not needed. Students can also use **Edit | Undo** to return the last piece, or they can press *Reset* to return all tiles to the stacks.

- In worksheet steps 8 and 9, there are two ways to factor each of these polynomials.

SUMMARIZE

Project the sketch. Expect to spend about 5 minutes.

9. Gather the class. Students should have their worksheets with them. Ask, *Today you factored polynomials. What does it mean to factor a polynomial?* Establish as a class that factoring is the opposite process of multiplying. You start with the product and need to find two factors that, when multiplied, give the product.

10. *How can the area of a rectangle be written as a sum and a product?* Students should be able to identify that the area of a rectangle is the sum of the parts, which is equivalent to the product of the dimensions. Factoring can be represented by the multiplication equation *area = length × width* (or, *area = base × height*), where the area corresponds to the product and the dimensions correspond to the factors.

11. *Does the order of the factors change the product?* Students should be able to justify the commutative property of multiplication as evidenced by the conservation of the area tiles.

12. *How can we use the factors of the last term of the polynomial to make the rectangular arrangement?* Students should be able to identify a means of using the constant term factors to determine the grouping of the x tiles.

13. If time permits, discuss the Explore More. Let students show arrangements they made using the custom tools and how they adjust to the values of the sliders. You may consider giving all students the chance to work with the dynamic tile tools on another day.

14. You may wish to have students respond individually in writing to this prompt: *Explain how to find the factors of a rectangle whose area tiles are $x^2 + 8x + 7$. Explain why this polynomial has only one pair of factors. Give an example of a rectangle that has more than one factor pair.*

EXTEND

Ask, *Will this factoring process work for polynomials that have more than one x^2? For example, what are the factors of $2x^2 + 3x + 1$?* $[(2x + 1)$ and $(x + 1)]$. Create more challenging factoring problems, such as $3x^2 + 20x + 12 = (3x + 2)(x + 6)$. Remind students that they can work this process in reverse. For instance, you might first ask, *What is the product of $(y + 1)(2x + 3)$?*

ANSWERS

Get Ready

A. 1, 6 and 2, 3

B.

The 6 should be placed inside because it is counting the interior squares.

C. Area = 12.7

Exploring Expressions and Equations in Grades 6–8 with The Geometer's Sketchpad
© 2012 Key Curriculum Press

D.

Type	Quantity
1	8
x	4
y	6
xy	1
x^2	0
y^2	1

E. Area is $4x + xy + 6y + y^2 + 8$; dimensions are $(y + 4)$ and $(y + x + 2)$.

F. $x^2 + 7x + 10$

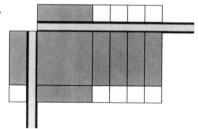

Dimensions are $(x + 2)$ and $(x + 5)$.

Explore

4.

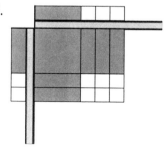

$x^2 + 5x + 4 = (x + 4)(x + 1)$

5.

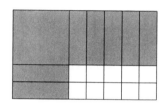

$x^2 + 5x + 6 = (x + 3)(x + 2)$

6.

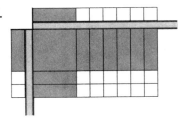

$$x^2 + 8x + 12 = (x + 6)(x + 2)$$

7.

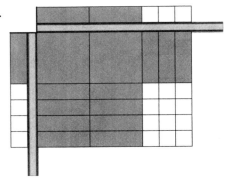

$$2x^2 + 11x + 12 = (2x + 3)(x + 4)$$

8.

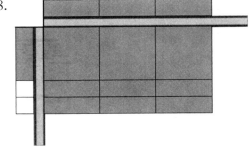

$$3x^2 + 6x = (3x)(x + 2) \text{ (or, } 3x^2 + 6x = (x)(3x + 6))$$

9.

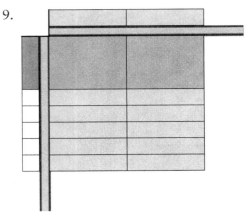

$$2xy + 10y = (2y)(x + 5) \text{ (or, } 2xy + 10y = (y)(2x + 10))$$

Exploring Expressions and Equations in Grades 6–8 with The Geometer's Sketchpad

10.

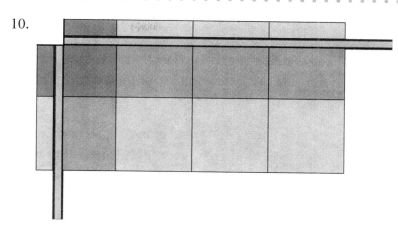

$$x^2 + 4xy + 3y^2 = (x + y)(x + 3y)$$

11.

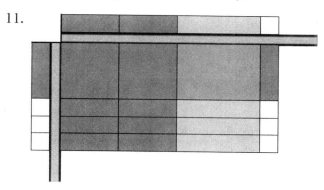

$$2x^2 + xy + 7x + 3y + 3 = (2x + y + 1)(x + 3)$$

12. *Note:* Numbers for vertical and horizontal x tiles can be reversed.

Problem	Polynomial	Number of Vertical x Tiles	Number of Horizontal x Tiles
Step 4	$x^2 + 5x + 4$	4	1
Step 6	$x^2 + 5x + 6$	3	2
Step 7	$x^2 + 8x + 12$	6	2

13. The numbers of vertical and horizontal tiles are equal to the constants of the factored expression.

14. $x^2 + 7x + 12 = (x + 3)(x + 4)$

15. The coefficient b is equal to the sum of the constants of the factored expression. The constant c is equal to the product of the constants in the factored expression.

Tiling Rectangles

Name:

GET READY

A. Factoring means finding the multipliers that make the product. The factor pairs of 4 are 1, 4 and 2, 2. What are the factor pairs of 6?

B. For each of the factor pairs you listed above, draw and label a rectangle that has the same dimensions as your factor pair. Where should the value 6 be placed for all of your rectangles? Why?

C. Notice that any number will have a factor pair that includes the number 1. Does the value 1 show up in the area? Find the area of the rectangle.

```
          12.7
┌──────────────────────────┐
│                          │ 1
└──────────────────────────┘
```

D. List the number of each type of tile in the arrangement.

Type	Quantity
1	
x	
y	
xy	
x^2	
y^2	

E. What is the area of the arrangement above? What are the dimensions?

F. What polynomial does the arrangement shown here represent? Can you arrange the pieces into a rectangle? If so, what are its dimensions?

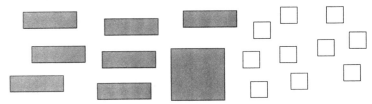

Exploring Expressions and Equations in Grades 6–8 with The Geometer's Sketchpad
© 2012 Key Curriculum Press

Tiling Rectangles

 Name:

In this activity you'll use algebra tiles to make rectangles and find their dimensions as a model for factoring polynomials.

EXPLORE

1. Open **Tiling Rectangles.gsp** and go to page "Frame."

2. Drag tiles from the stacks to the inside of the frame to represent the polynomial $x^2 + 5x + 4$. Arrange the tiles into a rectangle. You can exchange horizontal and vertical x tiles as needed.

3. Drag tiles to the outside of the frame to represent the dimensions of the rectangle.

4. Sketch the finished arrangement and express it as an area equation.

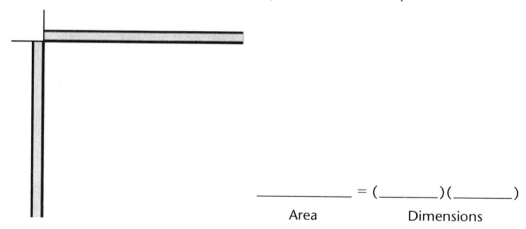

_____ = (_____)(_____)
   Area   Dimensions

5. Press *Reset*. Repeat steps 2–4 for the polynomial $x^2 + 5x + 6$.

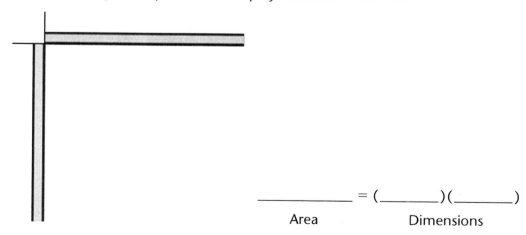

_____ = (_____)(_____)
   Area   Dimensions

6. Factor $x^2 + 8x + 12$. Sketch the finished arrangement.

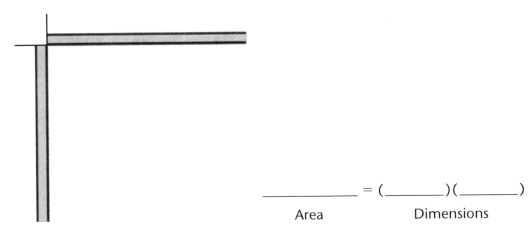

$$\underline{\hspace{3cm}} = (\underline{\hspace{2cm}})(\underline{\hspace{2cm}})$$
Area   Dimensions

7. Factor $2x^2 + 11x + 12$. Sketch the finished arrangement.

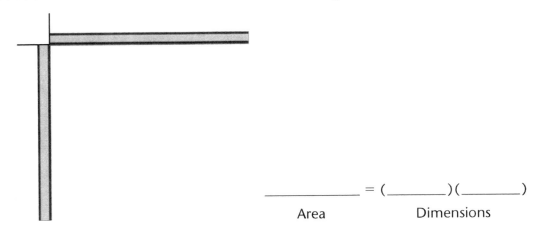

$$\underline{\hspace{3cm}} = (\underline{\hspace{2cm}})(\underline{\hspace{2cm}})$$
Area   Dimensions

8. Factor $3x^2 + 6x$. Sketch the finished arrangement.

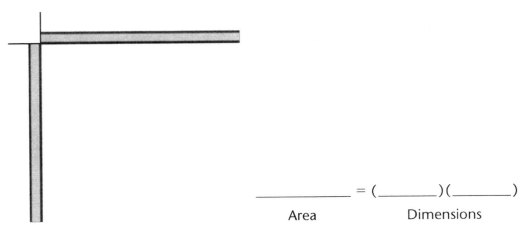

$$\underline{\hspace{3cm}} = (\underline{\hspace{2cm}})(\underline{\hspace{2cm}})$$
Area   Dimensions

Exploring Expressions and Equations in Grades 6–8 with The Geometer's Sketchpad
© 2012 Key Curriculum Press

Tiling Rectangles

continued

9. Factor $2xy + 10y$. Sketch the finished arrangement.

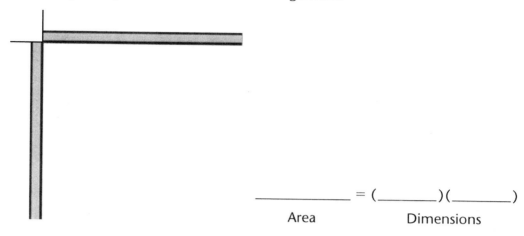

$$\underline{\hspace{2cm}} = (\underline{\hspace{1.5cm}})(\underline{\hspace{1.5cm}})$$
Area   Dimensions

10. Factor $x^2 + 4xy + 3y^2$. Sketch the finished arrangement.

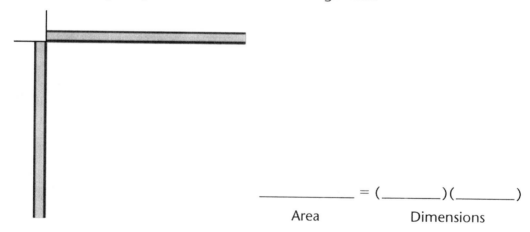

$$\underline{\hspace{2cm}} = (\underline{\hspace{1.5cm}})(\underline{\hspace{1.5cm}})$$
Area   Dimensions

11. Factor $2x^2 + xy + 7x + 3y + 3$. Sketch the finished arrangement.

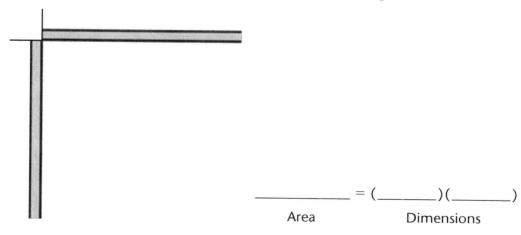

$$\underline{\hspace{2cm}} = (\underline{\hspace{1.5cm}})(\underline{\hspace{1.5cm}})$$
Area   Dimensions

12. Look back at steps 4, 5, and 6. In each of these problems, there is only one x^2 tile, and some number of x tiles and 1 tiles.

    For each problem, list how many x tiles are vertical and how many are horizontal in the rectangle inside the frame.

Problem	Polynomial	Number of Vertical x Tiles	Number of Horizontal x Tiles
Step 4	$x^2 + 5x + 4$		
Step 5	$x^2 + 5x + 6$		
Step 6	$x^2 + 8x + 12$		

13. How are the numbers of vertical and horizontal x tiles related to the dimensions of the rectangle?

14. Use your observation in step 13 to find the factors of $x^2 + 7x + 12$ without using the Sketchpad model.

15. For polynomials of the form $x^2 + bx + c$, how are the factors related to the values of b and c?

EXPLORE MORE

16. Go to page "Explore More." Use the dynamic tile tools in the Custom Tools menu to construct an arrangement. Choose a tool and click on an existing point in the sketch. There are three points you can start with, one for the inside of the frame and one for each dimension along the frame. To connect new tiles to existing tiles, move the pointer until a point on the existing tile becomes highlighted and then click.

    Drag the x and y sliders to see how the arrangement adjusts to their new values. If you've constructed your arrangement correctly, it will hold together when you drag the sliders.

Dynamic Tiles:
Evaluating Polynomial Expressions

This activity can be used by students whether or not they have had hands-on experience with algebra tiles. Most students, especially tactile learners, benefit from manipulating physical tiles. This activity, however, includes sliders that allow students to change the lengths of tiles with variable dimensions, which can't be modeled with physical tiles.

This activity can be completed in one extended 75-minute session or in two shorter sessions. If you wish to complete the activity in one 60-minute session, you might have the students do the Explore More on another day. The Explore More, which does not require the use of custom tools, can be done separately from the rest of the activity.

INTRODUCE

Project the sketch for viewing by the class. Expect to spend about 10 minutes.

1. Open **Dynamic Tiles.gsp.** Enlarge the document window to fill the screen.

2. If your students don't have prior hands-on experience with algebra tiles, go to page "Tiles." Make sure that the dimensions are showing. Explain that tiles are named after their areas. Make sure students can identify all their names. Then press *Show Areas.* Drag the *x* and *y* sliders to show that the area expressions match the dimensions for all values. If your students are already familiar with algebra tiles, you can skip this step.

3. Go to page "(A) xy." Say, *Today you're going to use Sketchpad to evaluate polynomial expressions. You'll use different values for the variables x and y to see how the expressions change. Before you start, I'll show you how to use the model.*

4. Make sure the grid is showing. Drag the sliders so that *x* = 8 and *y* = 6. Ask, *What is the area of the light blue rectangle?* [48] Drag the *x* slider to 3. *What is the area now?* [18] Drag the *y* slider to 4 and ask again. [12] Then press *Hide Grid.* *What are the names of these three tiles?* [Purple is *x*, green is *y*, and blue is *xy*.] Press *Show Names.*

5. Go to page "(B) x(2y)." Press *Show Variable Tiles* and ask for the names again. Ask, *How can I write an equation using these names to say that the length multiplied by the width is equal to the area?* Use a text caption, or use the board, to write the multiplication problem.

$$(x)(2y) = 2xy$$

6. Drag the sliders so that $x = 2$ and $y = 3$. Ask, ***What is the area of the rectangular region inside the frame when x is 2 and y is 3?*** [12] Let students respond, and then press *Show Unit Tiles.*

7. Drag the x slider so that $x = 5$. Explain how to use the custom tools. ***In some parts of this activity, you'll use custom tools to add more tiles to the sketch. If I want to use unit tiles to fill in the rectangle when x is 5 and y is 3, I click on the*** Custom ***tool icon and choose the*** 1 ***tool. Then I click on the sketch to create more unit tiles. In order for the new tiles to be connected to the tiles that are already there, I click on the points at the corners of the existing tiles. Notice how the point becomes highlighted when I'm at the right spot.*** Let students know that the custom tool remains active until they select the **Arrow** tool (or any other tool).

8. Model how to add the unit tiles, going from left to right as you add each row, until there are 30 tiles. Remind students that they can always use **Edit | Undo** if they make a mistake. You might mention the difference between the horizontal and vertical versions of the x, y, and xy tiles, and that x^2 means x^2.

9. If you want students to save their work, demonstrate choosing **File | Save As,** and let them know how to name and where to save their files.

DEVELOP

Expect students at computers to spend about 60 minutes.

10. Assign students to computers and tell them where to locate **Dynamic Tiles.gsp.** Distribute the worksheet. Tell students to work through step 30 and do the Explore More if they have time. Encourage students to ask their neighbors for help if they are having difficulty with Sketchpad.

11. Let pairs work at their own pace. As you circulate, here are some things to notice.

   • In worksheet step 1, when $x = 11$ and $y = 12$, students should be encouraged to use multiplication instead of counting units.

   • In worksheet step 11, help students as needed to correctly use the **1** custom tool. Move the tile into the frame, but do not click until the point on the neighboring tile is highlighted.

   • When transitioning from adding tiles with a custom tool to dragging the sliders, remind them to choose the **Arrow** tool (or any other tool) to deselect the custom tool.

Exploring Expressions and Equations in Grades 6–8 with The Geometer's Sketchpad
© 2012 Key Curriculum Press

- In worksheet step 13, the area formed by the unit tiles will not remain 42, but the students should witness that there are 18 unit tiles that are no longer part of the evaluation for $x = 4$ and $y = 3$.

- In worksheet step 18, students may need clarification on how to fill in the area in the frame using the custom tool.

- In worksheet step 21, students are asked to use unit tiles to fill in the area in the frame. These tiles will overlap with the y tiles and should be used to justify their algebraic evaluation answers.

- In worksheet step 24, students are told to use the **x (vertical)** tool to add a tile to the dimension on the left side along the outside of the frame. They must then figure out to use the **xy (y horizontal)** tool to fill in the rectangle inside the frame.

- In worksheet step 30, students are discouraged from using unit tiles to count the area. You might ask about the limitation of the counting, and make sure they are correctly evaluating expressions algebraically.

12. If students will save their work, remind them where to save it now.

SUMMARIZE

Expect to spend about 5 minutes.

13. Gather the class. Students should have their worksheets with them. If students have completed the Explore More, have them share their solutions from worksheet step 34. Otherwise, review steps 26–30. Ask for volunteers to demonstrate the algebraic steps that verify their answers. Discuss the two methods of evaluating these rectangular areas: the product of the dimensions, or the sum of the areas. Both require adherence to the order of operations.

EXTEND

What questions might you ask about evaluating expressions? Encourage curiosity. Here are some sample student queries.

What does it mean to be a variable?

Can x and y have values that are negative?

How many values are there for x and y?

Are there any tiles that never change?

What is the difference between an expression and an equation?

ANSWERS

1.

x Slider Value	y Slider Value	Area of Rectangle in Frame
8	6	48
2	6	12
3	2	6
11	12	132
x	y	xy

2.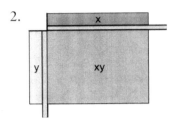

3. 12

4.

5. $2xy$

6. x and $2y$

7. $(x)(2y) = 2xy$

9. The tiles became longer vertically. The area is now more than 12.

11. 42

12. $7(2 \cdot 3) = 2 \cdot 7 \cdot 3$

   $7(6) = 14 \cdot 3$

   $42 = 42$

13. The area should now be less than 42 because the dimension tile on the left became shorter.

14. $4(2 \cdot 3) = 2 \cdot 4 \cdot 3$

   $4(6) = 8 \cdot 3$

   $24 = 24$

15. Answers will vary.

16. $2(2 \cdot 7.5) = 2 \cdot 2 \cdot 7.5$

$2(15) = 4 \cdot 7.5$

$30 = 30$

The light blue rectangle made up of two xy tiles has more area.

17.

18. $(3)(y) = 3y$

19. Nothing happens to the picture because there are no x-values.

20. The rectangular region gets bigger or smaller because all the y-values change together.

21. $3(3) = 9$

22. $3(6) = 18$

23. $3(9) = 27$

24. An xy tile (y horizontal) will fill in the area in the frame.

25.

26. $(x)(x + 2) = x^2 + 2x$

27. $2(2 + 2) = 2^2 + 2 \cdot 2$

$2(4) = 4 + 4$

$8 = 8$

28. $4(4 + 2) = 4^2 + 2 \cdot 4$

$4(6) = 16 + 8$

$24 = 24$

29. $5(5 + 2) = 5^2 + 2 \cdot 5$

$5(7) = 25 + 10$

$35 = 35$

30. $11(11 + 2) = 11^2 + 2 \cdot 11$

$11(13) = 121 + 22$

$143 = 143$

31.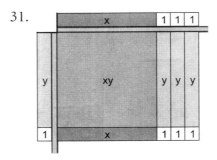

32. $(x + 3)(y + 1) = xy + x + 3y + 3$

33. $(9 + 3)(11 + 1) = 9 \cdot 11 + 9 + 3 \cdot 11 + 3$

$(12)(12) = 99 + 9 + 33 + 3$

$144 = 144$

34. Answers will vary. Sample solutions are shown in the tables.

12		16		30		36	
x	**y**	**x**	**y**	**x**	**y**	**x**	**y**
3	1	5	1	2	5	1	8
1	2	1	3	7	2	0	11
0	3	13	0	0	9	9	2
9	0	2	2.2	3	4	3	5
2	1.4	7	0.6	1	7.5	6	3

35. Answers will vary. Sample solutions are shown in the tables above.

Exploring Expressions and Equations in Grades 6–8 with The Geometer's Sketchpad
© 2012 Key Curriculum Press

Dynamic Tiles

In this activity you will use algebra tiles to evaluate polynomial expressions. See whether you can identify different methods for finding the area in the frame.

EXPLORE

1. Open **Dynamic Tiles.gsp** and go to page "(A) xy." Drag the *x* and *y* sliders to the values shown in the table and find the area of the light blue rectangle inside the frame.

x Slider Value	*y* Slider Value	Area of Rectangle in Frame
8	6	
2	6	
3	2	
11	12	
x	*y*	

2. Press *Hide Grid.* Draw a diagram of the tiles inside the frame and along the outside of the frame. Label each tile with its name. Then press *Show Labels* to check.

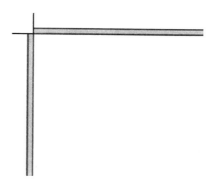

3. Go to page "(B) x(2y)." If needed, drag the *x* slider to 2 and the *y* slider to 3. What is the area of the yellow rectangle in the frame?

4. Press *Show Variable Tiles.* Draw a diagram of all the tiles and label them.

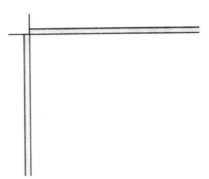

5. What is the area of the rectangle inside the frame?

6. What are the dimensions (length and width) represented by tiles along the outside of the frame?

7. Write the area equation shown by your diagram.

   (_____) (_____) = _____

8. To evaluate the expressions on both sides of the area equation, you replace the variables with their values and simplify. The diagram in step 4 shows that $x(2y) = 2xy$. In step 3, when $x = 2$ and $y = 3$,

$$2(2 \cdot 3) = 2 \cdot 2 \cdot 3$$

$$2 \cdot (6) = 4 \cdot 3$$

$$12 = 12$$

9. Change the *x*-value to 7. How did the tiles along the outside of the frame change? Is the area inside the frame more or less than 12?

10. Press *Show Unit Tiles.* Now you'll add unit tiles until you fill in the area of the rectangle when $x = 7$ and $y = 3$.

 11. Choose the **1** tool from the Custom Tools menu. Click the points along the bottom of the yellow squares starting from the left. Add enough unit tiles to fill in the area in the frame. How much area is in the frame?

12. Follow the process shown in step 8 to evaluate both sides of the equation when $x = 7$ and $y = 3$.

$$x(2y) = 2xy$$

$$\underline{}(2 \cdot \underline{}) = 2 \cdot \underline{} \cdot \underline{}$$

$$\underline{}(\underline{}) = \underline{} \cdot \underline{}$$

$$\underline{} = \underline{}$$

 13. Change the *x*-value to 4. Should the area determined by the new dimensions be more or less than 42? Why?

14. Evaluate both sides of the equation when $x = 4$ and $y = 3$.

15. Press *Show Variable Tiles*. Change the *x*-value to 2 and the *y*-value to 7.5. You should now see two rectangles, a light blue one made up of two *xy* tiles, and a yellow one made up of the unit tiles you added in step 11. Which one do you think has more area?

16. Find out by evaluating both sides of the equation $x(2y) = 2xy$ when $x = 2$ and $y = 7.5$.

17. Go to page "(C) 3*y*." Draw a diagram of the tiles along the outside of the frame and label them.

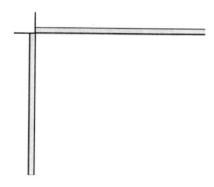

18. Choose the **y (horizontal)** tool. Fill the area inside the frame with *y* tiles. Add them to your diagram above and label them. Write the area equation shown by your diagram.

    (_____) (_____) = _____

19. Drag the *x* slider. What happens to the sketch? Why?

20. Drag the *y* slider. What happens to your sketch? Why?

21. Change the value of *y* to 3. Use the **1** tool to fill in the area inside the frame with unit tiles. Evaluate the expression 3*y* when *y* = 3.

22. Change the value of *y* to 6. Use the **1** tool to fill in the area inside the frame with unit tiles. Evaluate the expression 3*y* when *y* = 6.

23. Change the value of *y* to 9. Use the **1** tool to fill in the area inside the frame with unit tiles. Evaluate the expression 3*y* when *y* = 9.

24. Choose the **x (vertical)** tool and add an *x* tile to the dimension on the left side along the outside of the frame. Without using unit tiles, which tile will complete the rectangle in the frame? Use the appropriate custom tool and complete the rectangle in the frame.

25. Go to page "(D) x(x+2)." Draw a diagram of the tiles along the outside of the frame and label them.

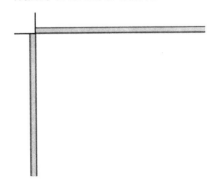

Exploring Expressions and Equations in Grades 6–8 with The Geometer's Sketchpad
© 2012 Key Curriculum Press

26. Fill in the area inside the frame without using any unit tiles. Choose the appropriate tools from the Custom Tools menu and add them to your sketch. Add them to your diagram above and label them. Write the area equation shown by your diagram.

(_____) (_____) = _____

27. Set x equal to 2. Fill in the area frame with unit tiles to count the area. Evaluate the expressions on both sides of the area equation for $x = 2$.

$$x(x + 2) = x^2 + 2x$$
$$\underline{\ }(\underline{\ } + 2) = \underline{\ }^2 + 2 \cdot \underline{\ }$$
$$\underline{\ }(\underline{\ }) = \underline{\ } + \underline{\ }$$
$$\underline{\ } = \underline{\ }$$

28. Repeat step 27 for $x = 4$.

29. Repeat step 27 for $x = 5$.

30. Repeat step 27 for $x = 11$ (but don't add tiles anymore!).

EXPLORE MORE

31. Go to page "(E) $(x+3)(y+1)$." Draw a diagram of all the tiles and label them.

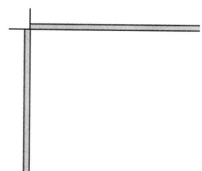

32. Write the area equation shown by your diagram.

    (_____)(_____) = _____

33. Drag the sliders so $x = 9$ and $y = 11$. Evaluate the expressions on both sides of the area equation in step 32 for $x = 9$ and $y = 11$. Then press *Show Grid* to compare your solution with the diagram.

    (_____)(_____) = _____

    (___)(___) = _____

    ____ = ____

34. Keep the grid showing. Use the x and y sliders to find the area in each table. There are many combinations of x and y for each area, so find as many different values as you can. Be prepared to verify your values of x and y using the method shown in step 33.

12		16		30		36	
x	y	x	y	x	y	x	y
3	1		1				
	2	1					
0			0				
	0						

35. In the tables above, use values of x and y that are not integers. Use calculations to check the estimated values from the sketch. How many values can you find?

Exploring Expressions and Equations in Grades 6–8 with The Geometer's Sketchpad
© 2012 Key Curriculum Press

Rooting for Roots:
Factoring and Graphing Quadratics

ACTIVITY NOTES

INTRODUCE

Project the sketch for viewing by the class. Expect to spend about 10 minutes.

1. Open **Rooting for Roots.gsp.** Go to page "Factoring."

2. Explain, *Today you'll look at how an area model can help you predict what the graph of a quadratic equation will look like. You'll start by exploring this area model in Sketchpad. You see both a quadratic expression and its area model. Can you interpret this area model? How do the four different rectangles relate to the expression?* Give students the opportunity to interpret the four rectangles (x^2, $2x$, $3x$, 6) and to describe how the sum of their areas is equal to the area of the large rectangle.

3. *What's another way to write the total area of the rectangle?* Students should see that the area can also be represented by the product $(x + 2)(x + 3)$. Tell students that this product of two binomials is the *factored form* of the expression $x^2 + 5x + 6$.

4. *You can change the value of each of these terms and this will change the expression and the area model as well. For example, I'll increase the horizontal term to 4.* Model how to change the value by double-clicking it and entering a new value or by selecting it and using the $+$ and $-$ keys on the keyboard.

5. *What should happen if I change the horizontal term to -3?* Ask students to write down both the product of the binomials and the expression. *This is how Sketchpad represents the corresponding area model. The areas of rectangles that are shaded black are subtracted from the sum of the other areas. Predict what will happen if I now change the vertical term to -2.* Students should be able to predict that the upper-left rectangle will be black too. If needed, ask them what will happen to the upper-right rectangle also. *The first expression we started with was $x^2 + 5x + 6$ and this one is $x^2 - 5x + 6$. We have created two expressions that have the same constant term of 6.*

DEVELOP

Expect students at computers to spend about 25 minutes.

6. Assign students to computers and tell them where to locate **Rooting for Roots.gsp.** Distribute the worksheet. Tell students to work through step 3.

7. Let pairs work at their own pace. As you circulate, here are some things to notice.

   - For worksheet steps 2a and 2b, there are several solutions. Ask students to choose one solution. The same is true for worksheet steps 3a and 3b.

   - Some students may still be struggling with the area model. Help them articulate the connection between the black and white regions and the associated product of binomials.

   - Encourage students to write their solutions both in product form and in expanded form.

8. Gather students together and compare their solutions to worksheet steps 2 and 3. Students will have a wide variety of answers for parts a and b, but should converge on part c.

9. Ask students to work through step 7 and do the Explore More if they have time. As you circulate, here are some things to notice.

   - Help students organize their sketches by moving the graphs out of the way of the area model and hiding the grid lines.

   - Remind students to click on the parameters when they are creating their graphs (instead of typing the values directly).

   - Also remind students how to change the parameters in order to change the graph.

SUMMARIZE

Project the sketch. Expect to spend about 10 minutes.

10. Open **Rooting for Roots.gsp** and go to page "Graphing." Use it to support the discussion. Begin by asking students to describe the differences between the two forms of quadratic equations they worked with in this activity (the expression in its expanded form and the factored form). Invite them to explain how the two forms are related to one another and why the factored form is helpful in predicting the corresponding graph.

11. Ask students to generate four quadratic equations that have a constant term of 18.

12. Ask students to generate examples of quadratic equations that have "special" characteristics, such as only one root, symmetric roots, and no roots.

Exploring Expressions and Equations in Grades 6–8 with The Geometer's Sketchpad
© 2012 Key Curriculum Press

ACTIVITY NOTES

EXTEND

1. Edit the function to create the graph of $f(x) = -x^2 + (m + n) \cdot x + m \cdot n$. Students should be able to see right away that the parabola now has a maximum instead of a minimum. Ask students whether this change has any effect on the relationship between the roots of the parabola and the m and n terms. Change the coefficient of x^2 to other values.

2. ***What other questions might you ask about this area model or the graphs of quadratic equations?*** Encourage all inquiry. Here are some ideas students might suggest.

    How else can you represent the negative length of negative area?

    What if you switched horizontal and vertical terms?

    How does the vertex of the parabola move if you animate the values of m or n?

    How can you find the coordinates of the vertex of a parabola?

ANSWERS

1. Students should be able to describe how the product of the two terms equals the constant term and their sum equals the coefficient of x.

2. a. Answers will vary. The product of the terms must be 12.
   b. Answers will vary. The product of the terms must be 12, and the horizontal term can be $-1, -2, -3, -4, -6,$ or -12.
   c. The vertical term is -12 and the horizontal term is -1 (or vice versa).

3. a. Answers will vary. The sum of the terms must be 8.
   b. Answers will vary. The sum of the terms must be 8 and the vertical term must be negative.
   c. The horizontal term is 11 and the vertical term is -3.

5. The graph crosses the x-axis at $-n$ and $-m$.

6. a. The parabola touches the x-axis at only one point.
   b. The parabola is symmetric about the y-axis.
   c. $f(x) = x^2 + 13x + 12$ and $f(x) = x^2 + -13x + 12$ are reflections across the y-axis, as are $f(x) = x^2 + 8x + 12$ and $f(x) = x^2 + -8x + 12$, and $f(x) = x^2 + 7x + 12$ and $f(x) = x^2 + -7x + 12$. All of the parabolas have a minimum whose x-coordinate is halfway between n and m.

7. a. $f(x) = x^2 + 4x$

   b. $f(x) = x^2 - 9$

   c. $f(x) = x^2 - 9x + 18$

8. Answers will vary. Sample answer: $f(x) = x^2 + 2$

Rooting for Roots

 Name:

In this activity you'll investigate how changing the graph of a quadratic equation is related to its roots, which in turn are related to the factored form of the quadratic equation.

EXPLORE

1. Open **Rooting for Roots.gsp.** Go to page "Factoring." You'll see an area model that represents the product of two binomials.

   If needed, change the vertical term to 2 and the horizontal term to 3.

   The area model represents $(x + 2)(x + 3)$. Change the values of both the horizontal and vertical terms, and describe how these terms relate to the quadratic expression at the top of the sketch.

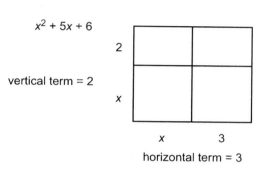

2. In the expression $x^2 + bx + c$, the term c is called the *constant* term. Change the values of the vertical and horizontal terms to

   a. Find a horizontal and vertical term in which $c = 12$.

   b. Find a horizontal and vertical term in which $c = 12$ and the horizontal term is negative.

   c. Find a horizontal and vertical term in which $c = 12$, the horizontal term is negative, and $b = -13$ (b is the coefficient of x).

3. Change the values of the vertical and horizontal terms to

   a. Find a horizontal and vertical term in which $b = 8$.

   b. Find a horizontal and vertical term in which $b = 8$ and the vertical term is negative.

   c. Find a horizontal and vertical term in which $b = 8$, the vertical term is negative, and $c = -33$.

4. Go to page "Graphing." You will use the values m and n (the vertical and horizontal terms in the area model) to create a graph.

   Choose **Graph│Plot New Function.**

To enter the function $f(x) = x^2 + (m + n) \cdot x + m \cdot n$, enter in order:

$x \wedge 2 + (\boldsymbol{m} + \boldsymbol{n}) * x + \boldsymbol{m} * \boldsymbol{n}$

Enter the bold letters by clicking the red value in the sketch. Click **OK**.

Choose **Graph│Hide Grid.**

Your graph should look something like this.

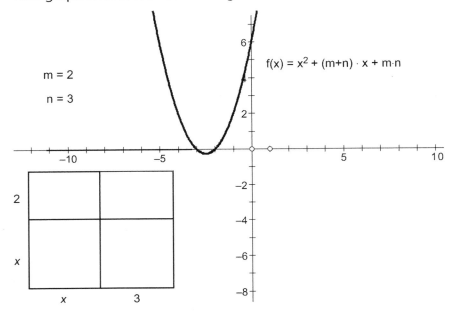

5. Describe the values at which the graph crosses the *x*-axis. How are they related to *m* and *n*?

6. Change the values of *m* and *n* (these are called the *roots* of the quadratic equation) and observe the effect on the graph you created.

   a. What happens when *m* and *n* have the same value?

   b. What happens when $n = -m$?

   c. Find all the possible graphs that have a constant term of 12 and write down their equations. How do they compare?

Exploring Expressions and Equations in Grades 6–8 with The Geometer's Sketchpad
© 2012 Key Curriculum Press

7. For each of the following graphs, write the corresponding equation. Use Sketchpad to verify your solution.

a.

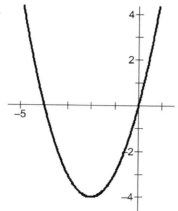

b.

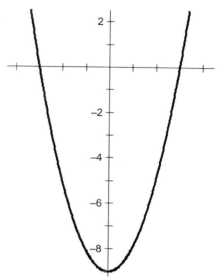

c.

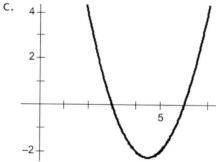

EXPLORE MORE

8. Use **Graph | Plot New Function** to generate three quadratic equations that do not have any roots.